D1277104

# Liquid Matter

Joseph A. Angelo, Jr.

*This book is dedicated to the faculty, staff, and students at Regis High School (New York City)— especially my fellow students in the class of 1961.*

**LIQUID MATTER**



Facts On File, Inc.
An imprint of Infobase Learning
132 West 31st Street
New York NY 10001

**Library of Congress Cataloging-in-Publication Data**
Angelo, Joseph A.
Liquid matter / Joseph A. Angelo, Jr.
p. cm.—(States of matter)
Includes bibliographical references and index.
ISBN 978-0-8160-7608-6
1. Liquids—Popular works. 2. Matter—Properties—Popular works. 3. Fluid dynamics—Popular works. 4. Water and civilization—Popular works. I. Title.
QC145.2.A54 2011
530.4'2—dc22 2010024865

Facts On File books are available at special discounts when purchased in bulk quantities for businesses, associations, institutions, or sales promotions. Please call our Special Sales Department in New York at (212) 967-8800 or (800) 322-8755.

You can find Facts On File on the World Wide Web at http://www.factsonfile.com



Text design by Annie O'Donnell
Composition by Hermitage Publishing Services
Illustrations by Sholto Ainslie
Photo research by the Author
Cover printed by Yurchak Printing, Inc., Landisville, Pa.
Book printed and bound by Yurchak Printing, Inc., Landisville, Pa.
Date printed: April 2011
Printed in the United States of America

10 9 8 7 6 5 4 3 2 1

This book is printed on acid-free paper.

# Contents

# Preface

*The unleashed power of the atom has changed everything save our modes of thinking.*

—Albert Einstein

**Humankind's global civilization relies upon a family of advanced** technologies that allow people to perform clever manipulations of matter and energy in a variety of interesting ways. Contemporary matter manipulations hold out the promise of a golden era for humankind—an era in which most people are free from the threat of such natural perils as thirst, starvation, and disease. But matter manipulations, if performed unwisely or improperly on a large scale, can also have an apocalyptic impact. History is filled with stories of ancient societies that collapsed because local material resources were overexploited or unwisely used. In the extreme, any similar follies by people on a global scale during this century could imperil not only the human species but all life on Earth.

Despite the importance of intelligent stewardship of Earth's resources, many people lack sufficient appreciation for how matter influences their daily lives. The overarching goal of States of Matter is to explain the important role matter plays throughout the entire domain of nature—both here on Earth and everywhere in the universe. The comprehensive multivolume set is designed to raise and answer intriguing questions and to help readers understand matter in all its interesting states and forms—from common to exotic, from abundant to scarce, from here on Earth to the fringes of the observable universe.

The subject of matter is filled with intriguing mysteries and paradoxes. Take two highly flammable gases, hydrogen ($H_2$) and oxygen ($O_2$), carefully combine them, add a spark, and suddenly an exothermic reaction takes place yielding not only energy but also an interesting new substance called water ($H_2O$). Water is an excellent substance to quench a fire, but it is also an incredibly intriguing material that is necessary for all life here on Earth—and probably elsewhere in the universe.

Matter is all around us and involves everything tangible a person sees, feels, and touches. The flow of water throughout Earth's biosphere, the air people breathe, and the ground they stand on are examples of the most commonly encountered states of matter. This daily personal encounter with matter in its liquid, gaseous, and solid states has intrigued human beings from the dawn of history. One early line of inquiry concerning the science of matter (that is, *matter science*) resulted in the classic earth, air, water, and fire elemental philosophy of the ancient Greeks. This early theory of matter trickled down through history and essentially ruled Western thought until the Scientific Revolution.

It was not until the late 16th century and the start of the Scientific Revolution that the true nature of matter and its relationship with energy began to emerge. People started to quantify the properties of matter and to discover a series of interesting relationships through carefully performed and well-documented experiments. Speculation, philosophical conjecture, and alchemy gave way to the scientific method, with its organized investigation of the material world and natural phenomena.

Collectively, the story of this magnificent intellectual unfolding represents one of the great cultural legacies in human history—comparable to the control of fire and the invention of the alphabet. The intellectual curiosity and hard work of the early scientists throughout the Scientific Revolution set the human race on a trajectory of discovery, a trajectory that not only enabled today's global civilization but also opened up the entire universe to understanding and exploration.

In a curious historical paradox, most early peoples, including the ancient Greeks, knew a number of fundamental facts about matter (in its solid, liquid, and gaseous states), but these same peoples generally made surprisingly little scientific progress toward unraveling matter's inner mysteries. The art of metallurgy, for example, was developed some 4,000 to 5,000 years ago on an essentially trial-and-error basis, thrusting early civilizations around the Mediterranean Sea into first the Bronze Age and later the Iron Age. Better weapons (such as metal swords and shields) were the primary social catalyst for technical progress, yet the periodic table of chemical elements (of which metals represent the majority of entries) was not envisioned until the 19th century.

Starting in the late 16th century, inquisitive individuals, such as the Italian scientist Galileo Galilei, performed careful observations and measurements to support more organized inquiries into the workings of the natural world. As a consequence of these observations and experiments, the nature of matter became better understood and better quantified. Scientists introduced the concepts of density, pressure, and temperature in their efforts to more consistently describe matter on a large (or macroscopic) scale. As instruments improved, scientists were able to make even better measurements, and soon matter became more clearly understood on both a macroscopic and microscopic scale. Starting in the 20th century, scientists began to observe and measure the long-hidden inner nature of matter on the atomic and subatomic scales.

Actually, intellectual inquiry into the microscopic nature of matter has its roots in ancient Greece. Not all ancient Greek philosophers were content with the prevailing earth-air-water-fire model of matter. About 450 B.C.E., a Greek philosopher named Leucippus and his more well-known student Democritus introduced the notion that all matter is actually composed of tiny solid particles, which are *atomos* (ατομος), or indivisible. Unfortunately, this brilliant insight into the natural order of things lay essentially unnoticed for centuries. In the early 1800s, a British schoolteacher named John Dalton began tinkering with mixtures of gases and made the daring assumption that a chemical element consisted of identical indestructible atoms. His efforts revived atomism. Several years later, the Italian scientist Amedeo Avogadro announced a remarkable hypothesis, a bold postulation that paved the way for the atomic theory of chemistry. Although this hypothesis was not widely accepted until the second half of the 19th century, it helped set the stage for the spectacular revolution in matter science that started as the 19th century rolled into the 20th.

What lay ahead was not just the development of an atomistic kinetic theory of matter, but the experimental discovery of electrons, radioactivity, the nuclear atom, protons, neutrons, and quarks. Not to be outdone by the nuclear scientists, who explored nature on the minutest scale, astrophysicists began describing exotic states of matter on the grandest of cosmic scales. The notion of degenerate matter appeared as well as the hypothesis that supermassive black holes lurked at the centers of most large galaxies after devouring the masses of millions of stars. Today, cosmologists and astrophysicists describe matter as being clumped into enormous clusters and superclusters of galaxies. The quest for these scientists is to explain how the observable universe, consisting of understandable forms of matter and energy, is also immersed in and influenced by mysterious forms of matter and energy, called dark matter and dark energy, respectively.

The study of matter stretches from prehistoric obsidian tools to contemporary research efforts in nanotechnology. States of Matter provides 9th- to 12th-grade audiences with an exciting and unparalleled adventure into the physical realm and applications of matter. This journey in search of the meaning of substance ranges from everyday "touch, feel, and see" items (such as steel, talc, concrete, water, and air) to the tiny, invisible atoms, molecules, and subatomic particles that govern the behavior and physical characteristics of every element, compound, and mixture, not only here on Earth, but everywhere in the universe.

Today, scientists recognize several other states of matter in addition to the solid, liquid, and gas states known to exist since ancient times. These include very hot plasmas and extremely cold Bose-Einstein condensates. Scientists also study very exotic forms of matter, such as liquid helium (which behaves as a superfluid does), superconductors, and quark-gluon plasmas. Astronomers and astrophysicists refer to degenerate matter when they discuss white dwarf stars and neutron stars. Other unusual forms of matter under investigation include antimatter and dark matter. Perhaps most challenging of all for scientists in this century is to grasp the true nature of dark energy and understand how it influences all matter in the universe. Using the national science education standards for 9th- to 12th-grade readers as an overarching guide, the States of Matter set provides a clear, carefully selected, well-integrated, and enjoyable treatment of these interesting concepts and topics.

The overall study of matter contains a significant amount of important scientific information that should attract a wide range of 9th- to 12th-grade readers. The broad subject of matter embraces essentially all fields of modern science and engineering, from aerodynamics and astronomy, to medicine and biology, to transportation and power generation, to the operation of Earth's amazing biosphere, to cosmology and the explosive start and evolution of the universe. Paying close attention to national science education standards and content guidelines, the author has prepared each book as a well-integrated, progressive treatment of one major aspect of this exciting and complex subject. Owing to the comprehensive coverage, full-color illustrations, and numerous informative sidebars, teachers will find the States of Matter to be of enormous value in supporting their science and mathematics curricula.

Specifically, States of Matter is a multivolume set that presents the discovery and use of matter and all its intriguing properties within the context of science as inquiry. For example, the reader will learn how the ideal gas law (sometimes called the ideal gas equation of state) did not happen overnight. Rather, it evolved slowly and was based on the inquisitiveness and careful observations of many scientists whose work spanned a period of about 100 years. Similarly, the ancient Greeks were puzzled by the electrostatic behavior of certain matter. However, it took several millennia until the quantified nature of electric charge was recognized. While Nobel Prize–winning British physicist Sir J. J. (Joseph John) Thomson was inquiring about the fundamental nature of electric charge in 1898, he discovered the first subatomic particle, which he called the electron. His work helped transform the understanding of matter and shaped the modern world. States of Matter contains numerous other examples of science as inquiry, examples strategically sprinkled throughout each volume to show how scientists used puzzling questions to guide their inquiries, design experiments, use available technology and mathematics to collect data, and then formulate hypotheses and models to explain these data.

States of Matter is a set that treats all aspects of physical science, including the structure of atoms, the structure and properties of matter, the nature of chemical reactions, the behavior of matter in motion and when forces are applied, the mass-energy conservation principle, the role of thermodynamic properties such as internal energy and entropy (disorder principle), and how matter and energy interact on various scales and levels in the physical universe.

The set also introduces readers to some of the more important solids in today's global civilization (such as carbon, concrete, coal, gold, copper, salt, aluminum, and iron). Likewise, important liquids (such as water, oil, blood, and milk) are treated. In addition to air (the most commonly encountered gas here on Earth), the reader will discover the unusual properties and interesting applications of other important gases, such as hydrogen, oxygen, carbon dioxide, nitrogen, xenon, krypton, and helium.

Each volume within the States of Matter set includes an index, an appendix with the latest version of the periodic table, a chronology of notable events, a glossary of significant terms and concepts, a helpful list of Internet resources, and an array of historical and current print sources for further research. Based on the current principles and standards in teaching mathematics and science, the States of Matter set is essential for readers who require information on all major topics in the science and application of matter.

# Acknowledgments

I wish to thank the public information and/or multimedia specialists at the U.S. Department of Energy (including those at DOE Headquarters and at all the national laboratories), the U.S. Department of Defense (including the individual armed services: U.S. Air Force, U.S. Army, U.S. Marines, and U.S. Navy), the National Institute of Standards and Technology (NIST) within the U.S. Department of Commerce, the U.S. Department of Agriculture, the National Aeronautics and Space Administration (NASA) (including its centers and astronomical observatory facilities), the National Oceanic and Atmospheric Administration (NOAA) of the U.S. Department of Commerce, and the U.S. Geological Survey (USGS) within the U.S. Department of the Interior for the generous supply of technical information and illustrations used in the preparation of this book set. Also recognized here are the efforts of Frank Darmstadt and other members of the Facts On File team, whose careful attention to detail helped transform an interesting concept into a polished, publishable product. The continued support of two other special people must be mentioned here. The first individual is my longtime personal physician, Dr. Charles S. Stewart III, M.D., whose medical skills allowed me to successfully work on this interesting project. The second individual is my wife, Joan, who for the past 45 years has provided the loving and supportive home environment so essential for the successful completion of any undertaking in life.

# Introduction

*"Water, water, every where,*
*And all the boards did shrink;*
*Water, water, every where,*
*Nor any drop to drink."*

—*The Rime of the Ancient Mariner*
(written in 1797–98 by Samuel Taylor Coleridge)

**The history of civilization is essentially the story of human beings** understanding and manipulating matter. This book presents many of the important discoveries that led to the scientific interpretation of matter in the liquid state. Readers will learn how the ability of human beings to relate the microscopic (atomic level) behavior of liquids to their readily observable macroscopic properties (such as density, pressure, temperature, and viscosity) helped transform the world.

Supported by a generous quantity of full-color illustrations and interesting sidebars, *Liquid Matter* describes the basic characteristics and properties of several important liquids—including water, blood, oil, and mercury. The three most familiar states of matter encountered on Earth are solid, liquid, and gas. Liquids are actually quite rare in nature considered on a cosmic scale. Stars are plasmas (the fourth state of matter). Giant planets are mainly gases, with frigid liquids that occur only at great depths below their cloud tops. Terrestrial (small) planets are generally solid and rocky, and the giant molecular clouds found in interstellar space contain mostly hydrogen and helium gas.

Earth is a pleasant exception to this apparent cosmic pattern. Humankind's home planet is just large enough, has a sufficiently dense atmosphere, and is located at a favorable enough distance from its parent star (the Sun) so that liquids, primarily water ($H_2O$), can exist and flow on its surface. The only other celestial body in the solar system where liquids are now known to exist and flow on the surface is Titan—the intriguing, cloud-enshrouded moon of the planet Saturn. However, the lakes, rivers, and shallow seas detected on the frigid surface of Titan (about −290°F

Earth imaged by NASA's *Galileo* spacecraft on December 11, 1990, when the spacecraft was about 1.3 million miles (2.1 million km) from the planet. South America and picturesque weather fronts in the South Atlantic Ocean appear in this image. *(NASA/JPL)*

[–179°C {94 K}]) consist of liquid hydrocarbons such as methane ($CH_4$) and not life-enabling water.

In general, a solid occupies a specific, fixed volume and retains its shape. A liquid also occupies a specific volume but is free to flow and assume the shape of the portion of the container it occupies. A gas has neither a definite shape nor a specific volume. Rather, it will quickly fill the entire volume of a closed container. Unlike solids and liquids, gases can be compressed easily. When temperatures get sufficiently high, another state of matter appears. Scientists call this fourth state of matter plasma. Finally, as the temperature gets very low and approaches absolute zero, scientists encounter a fifth state of matter called the Bose–Einstein condensate (BEC).

*Liquid Matter* presents the nature and scope of the science of fluids, highlights the more important scientific principles upon which the field is based, and identifies the wide range of applications that fluid science plays in almost all professional scientific and engineering fields. The role of liquids in important natural phenomena, such as Earth's hydrologic cycle, is also addressed.

The term *fluid* includes both liquids and gases. On Earth, water is the most commonly encountered liquid, and air is the most commonly encountered gas. This volume focuses its attention on liquids, although scientific principles common to all fluids (both liquids and gases) are included for clarity and continuity. Long before the Scientific Revolution that started in the middle of the 16th century in western Europe, ancient engineers and scientists, such as Archimedes of Syracuse (ca. 287–212 B.C.E.), examined the behavior of fluids and developed devices that harnessed, controlled, and applied natural fluid science phenomena such as flowing water and wind. Centuries later, during the Scientific Revolution, pioneering scientists such as Galileo Galilei (1564–1642), Sir

Isaac Newton (1642–1727), Evangelista Torricelli (1608–47), Blaise Pascal (1623–62), and Daniel Bernoulli (1700–82) began describing and predicting fluid behavior. Their technical efforts yielded important fluid science relationships for both liquids and gases. Early scientists and engineers used interesting experiments and mathematical relationships to help unlock nature's secrets. Their dedicated activities formed the broad field of fluid science (sometimes called fluid mechanics). Modern scientists and engineers continue this intellectual tradition but are now assisted by the sophisticated, computer-based methodology called computational fluid dynamics (CFD).

It is important to realize that the science of fluids forms the cornerstone of modern civilization. Such diverse activities as power generation, air transportation, maritime commerce, space exploration, reliable municipal water supplies, sanitary sewer systems, and modern medicine depend upon humankind's ability to understand and predict how liquids and gases behave under various physical conditions and circumstances.

Fluid science is the major branch of science that deals with the behavior of fluids (both gases and liquids) at rest (fluid statics) and in motion (fluid dynamics). This scientific field has many subdivisions and applications, including aerodynamics (the motion of gases including and especially air), hydrostatics (liquids, including water) at rest, and hydrodynamics (liquids in motion, including naturally flowing or artificially pumped water). *Liquid Matter* concentrates on the interesting phenomena and scientific principles associated with hydrostatics and hydrodynamics. Special attention is given to the more important liquids in a typical person's daily life—water, petroleum (oil), and bodily fluids (especially blood and urine). The metallic element mercury (Hg), which is liquid at room temperature (nominally 68°F [20°C]), is also examined in depth.

A fluid is a substance that, when in static equilibrium, cannot sustain a shear stress. A perfect or ideal fluid is one that has zero viscosity—that is, offers no resistance to shape change. Actual fluids only approximate this behavior. As previously mentioned, when scientists and engineers use the term *fluid,* they refer to both liquids and gases. One easy way to understand the difference is that a liquid will take the shape of and stay in an uncovered container, such as a bowl or coffee cup. In contrast, a gas easily takes the shape of the container but then escapes unless the container is tightly sealed. Viscosity is an important idea linked to internal fluid friction. This phenomenon is quite apparent when handling liquids. Water is

Lakes of liquid methane on Saturn's moon Titan, as radar-imaged by NASA's *Cassini* spacecraft on July 22, 2006 *(NASA/JPL/USGS)*

much less viscous than honey, even though both are liquids that flow at room temperature.

Although fluids are very important in daily life, most people do not have a good understanding of the scientific principles that govern the behavior of fluids. To help remedy this circumstance, *Liquid Matter* builds upon the three key ideas that form the overall architecture of fluid science. These important concepts are Newton's laws of motion, the continuity principle (indestructibility of flowing matter), and the conservation of energy. Scientists use other important concepts in their investigation of fluids. These additional notions include the idea of pressure (based upon Pascal's principle of hydrostatics), the concept of buoyancy (based upon Archimedes' famous principle), the perfect gas law, and the Bernoulli principle.

Historically, the scientific study of fluids concentrated first on the behavior of gases and then, as thermodynamics matured in the 19th century, incorporated the behavior of liquids under a variety of important physical conditions. Ancient people manipulated the flow of water, produced wine and beer, and used mercury to extract gold and silver from ore, but the scientific understanding of what actually happened when water vaporizes and becomes steam, how ocean currents govern the planet's climate, and how blood flows through the human body awaited sophisticated interpretations of liquid behavior.

*Liquid Matter* also examines the role of liquids in the context of the development of human civilization. This often ignored historic perspective helps highlight the importance of liquids in human life. One example will suffice here to illustrate this important point. Most people in developed countries simply flush a toilet without any knowledge of where the input water came from or where the outgoing waste stream is traveling to. The only time such people pay any attention to their toilets is when the convenient human waste removal devices become clogged and overflow. A similar "liquid crisis" occurs when the hot water supply runs out during a person's shower.

Modern water, sanitation, and hygiene systems represent one of humankind's great engineering achievements, yet, as discussed in this volume, there is still much room for improvement on a global basis. Safe drinking water, sanitation, and hygiene represent the three most important conditions for keeping communities healthy. According to the World Health Organization, unsafe drinking water coupled with a lack of basic sanitation kills at least 1.6 million children under the age of five each year. Furthermore, about 2.6 billion people around the world do not use a toilet—they defecate in open fields or other unsanitary places instead.

*Liquid Matter* provides numerous examples of how fluid science principles and concepts govern many fields, including astronomy, biology, medical sciences, chemistry, Earth science, meteorology, geology, oceanography, and physics. Fluid science forms an integral part of many branches of engineering, including aeronautics and astronautics, biological engineering, chemical engineering (including petroleum extraction and refining), civil engineering (including dam construction and calculating wind loads on tall buildings), environmental and sanitary engineering, industrial engineering (including food and beverage processing), mechanical engineering, nuclear engineering, and ocean engineering (including shipbuilding).

This book has been carefully designed to help any student or teacher who has an interest in the behavior and application of liquids discover what liquids are, how scientists measure and characterize them, and how the fascinating properties and characteristics of liquids influenced the course of human civilization. The back portion of the book contains an appendix with a contemporary periodic table, chronology, glossary, and array of historical and current sources for further research. These should prove especially helpful for readers who need additional information on specific terms, topics, and events.

The author has carefully prepared the material so that a person familiar with SI units (the international language of science) will have no difficulty understanding and enjoying its contents. The author also recognizes that there is a continuing need for some students and teachers in the United States to have units expressed in the American customary system of units. Wherever appropriate, both unit systems appear side by side. An editorial decision places the American customary units first, followed by the equivalent SI units in parentheses. This format does not imply the author's preference for American customary units over SI units. Rather, the author strongly encourages all readers to take advantage of the particular formatting arrangement to learn more about the important role that SI units play within the international scientific community.

CHAPTER 1

# Liquid Matter—An Initial Perspective

*"Παντα ρει (everything flows)"*
—Greek philosopher Heraclitus (ca. 535–475 B.C.E.)

**The ability of human beings to relate the microscopic (atomic level)** behavior of liquid matter to readily observable macroscopic properties (such as density, pressure, temperature, and viscosity) transformed science and engineering. Archimedes and other great engineers of antiquity used liquids effectively without possessing a detailed knowledge of their microscopic composition. The real breakthroughs in accurately describing the various properties of liquid matter started during the Scientific Revolution. This chapter explores how early Greek engineers and natural philosophers initiated the study of fluids. Their work, expanded and refined during the Scientific Revolution, exerted a dominant influence on the emergence of today's technology-enriched global civilization.

## RIVERS CRADLE EARLY CIVILIZATIONS

Flowing water served as the natural organizing principle for and the lifeblood of the first permanent human communities. Anthropologists sometimes refer to such early urban societies as hydraulic civilizations. In the ancient Middle East, the first civilizations arose along the Nile River (in

## THE FERTILE CRESCENT

Rain and snow that falls high in the mountains of southern Turkey eventually becomes the Tigris and Euphrates Rivers. The life-enabling water flows southward across the arid terrain of modern Turkey, Syria, and Iraq into the Persian Gulf. More than 8,000 years ago, the region between the Tigris and Euphrates Rivers, called the Fertile Crescent, gave rise to early human civilizations. Irrigation anchored the Neolithic Revolution in this region of the world and allowed its early peoples to transition from hunter-gatherers into farmers. Irrigation still plays a major role here in sustaining agriculture.

The three most notable ancient civilizations that arose in Mesopotamia were those of the Sumerians (ca. 3500–2000 B.C.E.), the Babylonians (ca. 1800–539 B.C.E.), and the Assyrians (ca. 1350–612 B.C.E.). Since nature did not provide a regular supply of water to the region (especially southern Mesopotamia), the use of irrigation was critical for successful agriculture and abundant crop yields—necessary conditions for cities to develop and civilizations to emerge. The ancient cities of Mesopotamia became centers of commerce, government, religious activities, and defense. The lack of natural barriers (such as impassable mountain ranges) made this rich agricultural region attractive to belligerent neighboring peoples. From the dawn of human history, Mesopotamia experienced numerous invasions and regional conflicts. In 689 B.C.E., the Assyrian king Sennacherib destroyed the city of Babylon and made Nineveh the capital city of the Assyrian Empire. Then, in 612 B.C.E., the Assyrian Empire collapsed due to invasions by the Medes (tribes from Persia) and the Babylonians. Finally, in 539 B.C.E., the Persians conquered the city of Babylon, causing the collapse of the Babylonian Empire. The presence of petroleum resources in this region has projected its chronic geopolitical turmoil into modern times.

The ancient city of Ashur is located on the Tigris River in the northern part of Mesopotamia about 175 miles (280 km) north of Baghdad, Iraq. Founded sometime in the third millennium B.C.E., Ashur served as the political and reli-

Egypt) and in Mesopotamia, a Greek word meaning "between rivers." The rivers referred to are the Tigris River and Euphrates River that now flow through modern Iraq. Historians call Mesopotamia the Fertile Crescent or the Cradle of Civilization. Early civilizations also appeared in the Indus River Valley of South Asia and in the Yellow River Valley of China. Despite numerous external influences and social transformations, these four

**In this 2008 photograph, a U.S. Army soldier stands guard while overlooking the Tigris River at the ancient Assyrian city of Ashur (now Qal'at Shergat, Iraq). Ashur is a United Nations–recognized World Heritage Site.** ***(Department of Defense)***

gious capital of the Assyrian Empire from the 14th to the 9th centuries B.C.E. The ancient city sits astride the ecological boundary between rain-fed agriculture and irrigation–supported agriculture. The Assyrians named the city after Assur, their major deity. As a religious center, the city contained temples for the worship of Assur and other deities, such as the goddess Ishtar. In 2003, the United Nations Educational, Scientific, and Cultural Organization (UNESCO) designated Ashur a World Heritage Site.

ancient civilizations have served as the basis of continuous cultural development in the same geographic locations throughout recorded history.

Anthropologists, archaeologists, and other scientists suggest that the ancient hydraulic civilizations shared the following common features: The inhabitants practiced sophisticated levels of agriculture, including irrigation and the domestication of animals; they learned to use metals and to make advanced types of pottery and bricks; they constructed cities; they learned how to store water and use natural bodies of surface water for transportation; they produced alcoholic beverages, such as beer and wine; they developed complex social structures, including class systems; and they developed various forms of writing.

While humankind existed long before these early civilizations appeared, the ability to write made recorded history possible. Anthropologists regard the use of a written language as a convenient division between prehistoric and historic societies. The time periods cited for the Neolithic (New Stone Age) Revolution in this section are only approximate and vary significantly from geographic region to geographic region.

The Neolithic Revolution was the incredibly important transition from hunting and gathering to agriculture. As the last ice age ended about 12,000 years ago, various prehistoric societies in the Middle East (the Nile Valley and Mesopotamia) and in parts of Asia independently began to adopt crop cultivation. Since farmers tend to stay in one place, the early peoples involved in the Neolithic Revolution began to establish permanent settlements. Historians often identify this period as the beginning of human civilization. The Latin word *civis* means "citizen" or a "person who inhabitants a city."

The Neolithic Revolution took place in the ancient Middle East starting about 10,000 B.C.E. and continued until approximately 3500 B.C.E. During this period, increasing numbers of people shifted their dependence for subsistence from hunting and gathering to crop cultivation and animal domestication. The growing surplus of food and the control of fire set the stage for the emergence of artisan industries, such as pottery making and metal processing. People also learned how to use water transport for marketing their surplus food and artisan-created goods. These developments gave rise to commodities merchants, who engaged in geographically dispersed trading activities.

Archaeologists suggest that human beings used the barter system at least 100,000 years ago. As trading activities increased within and between the early civilizations that emerged in the ancient Middle East and around the eastern part of the Mediterranean Sea, the need for a more efficient means of measuring goods and storing wealth also arose. In about 3000 B.C.E., the people of Mesopotamia began using a unit of weight called the *shekel* to identify a specific mass of barley (estimated to be about 0.022 lbm [0.010 kg]). The shekel became a generally accepted weight of convenience for trading other commodities, including copper, bronze, and silver. In about 650 B.C.E., the Lydians (an Iron Age people whose kingdom was located in the western portion of modern Turkey) invented money when they introduced officially stamped pieces of gold and silver that contained specific quantities of these precious metals. Money has been the lubricant of global economic activity ever since.

The ancient Egyptians pioneered the development of river craft, since the slow-flowing Nile River was ideal for transportation. This 1963 stamp from San Marino depicts an early Egyptian cargo ship designed to carry agricultural produce. *(Author)*

During the Neolithic period, ancient peoples developed better tools and invented a variety of simple machines. As populations swelled, these new tools and simple machines allowed early societies to use different materials to modify their surroundings to better suit survival and growth. The natural environment changed as simple dams, irrigation canals, roads, and walled villages appeared. The rise of prosperous ancient civilizations in Mesopotamia is directly attributed to skilled stewardship of water resources, especially the introduction of innovative irrigation techniques.

With each development came the need for better devices to perform difficult or monotonous tasks. Simple machines such as the wheel and axle and the inclined plane yielded major breakthroughs in the ability to perform work. Early waterwheels appeared in ancient Greece more than 2,000 years ago, and people used them to grind grain. Often, these simple, fluid-powered machines allowed fewer workers to accomplish the monotonous but important tasks previously assigned to a larger number of workers. Other early machines such as pulleys were typically not powered by flowing fluids. These devices required human labor (muscle power) or animal power to operate and generally needed human supervision to perform a task or to take advantage of some physical principle such as gravity. Gravity was instinctively perceived but not quantitatively understood. The quantitative understanding of materials and the physical principles that governed the performance of tools and machines did not take place until the Scientific Revolution. From the technology-rich perspective of the 21st century, it is difficult to appreciate that until the arrival of practical steam engines in the 18th century, flowing water or wind powered every machine device beyond those operated solely by human muscles or animal power.

With the transition to agriculture came food surpluses, the domestication of animals, the storage and channeling of water, the production of clothing, the growing use of metals, the expansion of trade, the further specialization of labor, and a variety of interesting social impacts. The rise of civilization during the Neolithic Revolution spawned the first governments (based on the need to organize human labor and focus wealth in the development of various public projects); the collection of taxes (to pay for government); the first organized military establishments (to protect people and their possessions); and the first schools (to pass knowledge and technical skills on to future generations in a more or less organized fashion).

When bountiful economic circumstances permitted, a few people were even able to earn a living by simply thinking or teaching. These individuals became the philosophers, mathematicians, and early scientists (natural philosophers), whose ideas and accumulated knowledge began to influence the overall direction and rate of human development. As discussed in the next section, the philosophers and mathematicians of ancient Greece exerted a tremendous influence on the overall trajectory of Western civilization.

## EARLY CONCEPTS OF MATTER

Thales of Miletus (ca. 624–545 B.C.E.), the ancient Greek philosopher, was the first European thinker to suggest a theory of matter. About 600 B.C.E., he postulated that all substances came from water and would eventually turn back into water. Thales may have been influenced in his thinking by the fact that water assumes all three commonly observed states of matter: solid (ice), liquid (water), and gas (steam or water vapor). The evaporation of water under the influence of fire or sunlight could have provided him the notion of matter recycling in nature.

Anaximenes (ca. 585–525 B.C.E.) was a later member of the school of philosophy at Miletus. He followed in the tradition of Thales by proposing one primal substance as the source of all other matter. For Anaximenes, that fundamental substance was air. In surviving portions of his writings, he suggested "Just as our soul being air holds us together, so do breath and air encompass the whole world." He proposed that cold and heat and moisture and motion made air visible. He further stated that when air was rarified, it became diluted and turned into fire. In contrast, the winds for Anaximenes represented condensed air. If the condensation process continued, the result was water; further condensation resulted in the primal substance (air) becoming earth and then stones.

The Greek pre-Socratic philosopher Empedocles (ca. 495–435 B.C.E.) lived in Acragas, a Greek colony in Sicily. In about 450 B.C.E., he wrote the poem *On Nature*. In this lengthy work (of which only fragments survive), he introduced a theory of the universe in which all matter is made up of four classical elements—earth, air, water, and fire—that periodically combine and separate under the influence of two opposing forces (love and strife). According to Empedocles, when combined, fire and earth produce dry conditions; earth blends with water to form cold; water combines with air to produce wet; and air and fire when combined form hot.

In about 430 B.C.E., another early Greek philosopher named Democritus (ca. 460–370 B.C.E.) elaborated upon the atomic theory of matter initially suggested by his teacher Leucippus (fifth century B.C.E.). Democritus emphasized that all things consist of changeless (eternal), indivisible, tiny pieces of matter called atoms (ατομος). According to Democritus, different materials consisted of different atoms, which interacted in certain ways to produce the particular properties of a specific material. Some types of solid matter consisted of atoms with hooks, so they could attach to each other. Other materials, like water, consisted of large, round atoms that moved smoothly past each other. What is remarkable about the ancient Greek theory of atomism is that it tried to explain the great diversity of matter found in nature with just a few basic ideas tucked into a relatively simple theoretical framework. Unfortunately, the atomistic theory of matter fell from favor when the more influential Greek philosopher Aristotle (384–322 B.C.E.) rejected the concept. It was not until the 19th century that the concept of atoms returned to scientific discussions about matter.

Around 340 B.C.E., Aristotle embraced and embellished the theory of matter originally proposed by Empedocles. Within Aristotelian cosmology, planet Earth is the center of the universe. Everything within Earth's sphere is composed of a combination of the four basic elements: earth, air, water, and fire. Aristotle further suggested that objects made of these basic four elements are subject to change and move in straight lines. However, heavenly bodies are not subject to change and move in circles. Aristotle also stated that beyond Earth's sphere exists a fifth basic element, which he called aether (αιθηρ), a pure form of air that is not subject to change. Finally, he suggested that he could analyze all material things in terms of their matter and their form (essence). Aristotle's ideas about the nature of matter and the structure of the universe dominated thinking in Europe for centuries until finally being displaced during the Scientific Revolution.

## WATER CLOCKS KEEP TIME IN THE ANCIENT WORLD

Other ancient Greeks did more than contemplate the nature of matter. As some the ancient world's more famous engineers, they learned through trial and error how to manipulate and use matter in a variety of intriguing ways. The spectacular devices and discoveries of Archimedes, the legendary Greek engineer, generally overshadow the technical accomplishments of another famous Greek inventor and engineer, Ctesibius of Alexandria (ca. 285–222 B.C.E.). Ctesibius made many contributions to the fledgling disciplines of pneumatics, hydraulics, mechanics, and machine design.

He published an important work entitled *On Pneumatics* in 255 B.C.E. in which he discussed the elasticity of the air and enumerated various applications of compressed air in such devices as pumps, musical instruments, and even an air-powered cannon. His efforts in this area represent the start of *pneumatics,* the branches of science and engineering that deal with mechanical motions induced by pressurized gases. Unfortunately, complete copies of this work along with other writings perished. About 15 years later, Ctesibius introduced a greatly improved clepsydra (water clock), which became the best timepiece in antiquity and remained unrivaled in accuracy until pendulum clocks appear in Europe in the 17th century. In the chaos that accompanied the collapse of the Roman Empire, great libraries such as the one in Alexandria (Egypt) were destroyed and their contents scattered or destroyed.

What is specifically known about the engineering accomplishments of Ctesibius comes from other Greek inventors and engineers, such as Heron of Alexandria (ca. 20–80 C.E.) and the first-century B.C.E. Roman military engineer and architect Marcus Vitruvius Pollio (ca. 80–15 B.C.E.). Ctesibius is credited with the invention of the siphon, a flexible tube that can intake and lift a liquid over a height (elevation) using the pressure of the atmosphere and then discharge the liquid at a lower height (elevation). Ctesibius recognized that to start the action of a siphon, the device must first be filled with liquid before it is carefully placed in position. He is also considered the creator of a small pipe organ (the hydraulis), which was supplied with air by a piston pump.

His greatest technical accomplishment was a vastly improved version of the water clock (clepsydra) of ancient Egypt. Mechanical clocks were developed in medieval Europe and were based on falling weights. These devices were more convenient than Ctesibius's clepsydra, but not as accurate. By the mid-17th century, the Dutch astronomer and physicist Christiaan Huygens (1629–95) introduced the pendulum clock. His device surpassed the accuracy of the water clock and ushered in a new era in timekeeping. In credit to Ctesibius, few human-made devices have so dominated an area of technology (here, horology, or the science of measuring time) for almost two millennia.

## ARCHIMEDES OF SYRACUSE

The Greek mathematician, inventor, and engineer Archimedes of Syracuse was one of the greatest technical minds in antiquity, if not all history. As a gifted mathematician, he perfected a method of integration that allowed him to find the surface areas and volumes of many bodies. This brilliant work anticipated by almost two millennia the independent codevelopment of the calculus by Sir Isaac Newton and Gottfried Wilhelm Leibniz (1646–1716) in the middle of the 17th century. In mechanics, Archimedes discovered fundamental theorems and physical relationships that described the center of gravity of plane figures and solids. These relationships lie at the very heart of modern mechanics and engineering dynamics.

He designed and constructed a variety of potent war machines in defense of his birth city of Syracuse against sieges by the Roman army during the Second Punic War. These military devices, not his brilliant mathematical contributions, made Archimedes famous in his own lifetime. Today, science historians consider the affable, absentminded

This 1983 stamp honors Archimedes and depicts the famous Archimedes' screw, which is still used in many parts of the world to distribute water into irrigated fields. *(Author)*

Greek genius mathematician comparable in brilliance to Newton, Leonhard Euler (1707–83), or Carl Friedrich Gauss (1777–1855).

Archimedes was born ca. 287 B.C.E. in the Greek city-state of Syracuse on the island of Sicily. His father was an astronomer named Phidias, about whom very little else is known. As a young man, Archimedes studied at the great library in Alexandria. Since he was a distant relative of Hieron II, king of Syracuse, Archimedes elected to return to his birth city and pursue his interests in mathematics, science, and mechanics. Although he personally regarded mathematics as a much higher level of activity than his efforts involving the invention of various mechanical devices, it was these engineering efforts and not his mathematics that earned him great fame during his lifetime.

In approximately 250 B.C.E., he designed an endless screw, later named the Archimedes' screw. This important fluid-moving device could efficiently remove water from the hold of a large sailing ship as well pump water from rivers and lakes to irrigate arid fields. The device consisted of a helical screw enclosed in a cylindrical casing that was open at both ends. (See illustration.) The lowest portion of the pumping device is placed into the source body of water. As the helical screw turns, it scoops up a quantity of water. Subsequent revolutions of the screw move that quantity of water to other (higher) threads. Each revolution also adds a new scoop of water to the lowest thread. As the screw turns, water travels from thread to thread until it reaches the top of the cylinder. There, it leaves the screw and pours into the intended destination.

At first, human or animal power turned the Archimedes' screw, then people learned how to harness wind power to operate this simple yet efficient device. The helical screw usually fits snugly inside its cylindrical case but does not necessarily make a watertight seal. Save for the lowest thread, any water that seeps between threads during pumping is simply captured by a lower tread and hoisted upward again. Originally designed to help Egyptian peasants draw water out of the Nile River to irrigate their fields, the Archimedes' screw quickly found use throughout the Mediterranean Basin and the Middle East. People still use the device to move water, especially in certain underdeveloped regions of Africa and Asia.

One of the most famous stories about Archimedes involves a challenge extended to him by Hieron II. The king of Syracuse wanted to determine whether a goldsmith made a requested crown by using the proper amount of pure gold, as instructed. There was some suspicion that the goldsmith had cheated the king by using a less expensive combination of silver and gold. Hieron II asked Archimedes to give him the right answer but without damaging the new crown in any way.

Archimedes thought for days about this problem. Then, as often happens to creative people, inspiration struck when least expected. As Archimedes stepped into a full bath, he observed the water spill over the sides. Immediately, he knew how to solve the intriguing problem. Jumping up out of the bath, he ran naked to the palace, enthusiastically shouting "Eureka!" This Greek exclamation means "I have found it!" Today, when an engineer or scientist experiences a similar intellectual breakthrough, they are said to have a "eureka moment."

What Archimedes had discovered in a flash of genius was the principle of buoyancy. Today, physicists refer to this phenomenon as the Archimedes principle. The principle states that any fluid applies a buoyant force to an object when that object is partially or completely immersed in it. The magnitude of this buoyant force equals the weight of the fluid the object displaces. Since the shape of an object is of no consequence, Archimedes was able to use this phenomenon to test the king's crown without damaging it. In antiquity (as now), silver was much cheaper than gold, so it was often used as a substitute for gold by unscrupulous goldsmiths and jewelry merchants. Since silver has a lower density than gold, a crown fraudulently pasted up with silver in its interior would be bulkier (that is, have more volume) than a crown of identical weight made of only pure gold.

If the volume of water displaced by the pure gold were equal to the volume of water displaced by the submerged crown, both could be assumed to have the same density and consist of identical material, namely pure gold. If the volume of water displaced by the crown were different than the volume of water displaced by an identical weight of pure gold, then there was some other metal, possibly lead or silver, in the crown along with the gold. Archimedes carefully tested the suspicious crown and observed that the king's new crown displaced more water than the same weight of pure gold. So he concluded that the king's crown contained both gold and some other, less dense metal, most likely silver. The unwise goldsmith, who tried to cheat King Hieron II, was executed.

Today, scientists consider the Archimedes principle one of the basic laws of hydrostatics. This principle states that any object, entirely or partially submerged in a fluid (liquid or gas), is buoyed up by a force equal to the weight of the fluid displaced by the object.

Archimedes developed a number of important fundamental machines, including the lever and the compound pulley. With respect to the lever principle, Archimedes is reputed to have said: "Give me a place to stand on and I can move the Earth." Again challenged by King Hieron II to move something very large, Archimedes developed a system of compound pulleys and levers. Then, according to legend, he pulled a fully loaded ship containing crew and cargo up out of the water and onto the shore with a

## TIP OF THE ICEBERG

An iceberg is a large mass of ice freely floating in the ocean after being calved (detached) from a glacier or large ice sheet. Icebergs provide a dramatic natural demonstration of the concept of buoyancy. Because ice is about 90 percent the density of liquid water, an iceberg floats in water exposing just 10 percent or so of its total volume above the surface. The other 90 percent of the iceberg remains hidden beneath the waves. An ice cube floating in a glass of water also demonstrates this physical principle. The most famous iceberg in modern history is the one that collided with and sank the Royal Mail Ship (RMS) *Titanic* in 1912. When people want to emphasize that what is seen is only a small part of what is really there, they often utter the phrase "tip of the iceberg."

**A pinnacle iceberg floating off the coastline of Elephant Island, Antarctica. The ice-covered island received its name from the elephant seals that often occupy its shores.** ***(NOAA)***

single rope. Archimedes conducted other studies of force and motion. He discovered that every rigid body has a center of gravity, a single point at which the force of gravity appears to act on the body.

Many of Archimedes' surviving documents portray his wide-ranging interest in engineering and machines. These surviving works include *Theory of Levers, On Floating Bodies, On the Method of Mechanical Theorems,* and *The Water Clock.* He wrote many engineering-themed works, which today are known only from cross-references and prefaces in surviving books. Some of his missing works include *On Odometers, Winches, Hydroscopes, Pneumatics, On Balances or Levers, Centers of Gravity, Elements of Mechanics, On Gravity and Buoyancy,* and *Burning by Mirror.*

In addition to his genius for engineering, Archimedes was an extremely gifted mathematician who solved many important mathematical problems. He made the most precise estimates of the value of $\pi$ (the ratio of a circle's circumference to its diameter) of his day. He was a prolific writer in the field of mathematics, and some of his most important surviving works include *On the Sphere and Cylinder, Measurement of the Circle, On Spirals, On Tangential Circles, On Triangles,* and *On Quadrangles.* Despite achieving great fame through his mechanical inventions, Archimedes preferred to delve into mathematical problems, often getting absorbed for days and becoming oblivious to the world around him.

During his lifetime, Rome and Carthage fought for control of the Mediterranean basin. This power struggle resulted in a number of bloody conflicts called the Punic Wars. Rome waged three wars against Carthage: the First Punic War (264–241 B.C.E.), the Second Punic War (218–201 B.C.E.), and the Third Punic War (149–146 B.C.E.). Carthage was defeated and totally destroyed in the Third Punic War, leaving Rome in complete control of the ancient world around the Mediterranean Sea.

Archimedes became famous as a result of the many machines he developed for the defense of Syracuse during the First and Second Punic Wars. Specifically, he designed a variety of intricate machines to repulse attackers. Historians often place his military machines into three basic categories. First, there were the Archimedes' claws—cranes that could lift enemy ships up out of the water and smash them against the rocks. Next, there were a variety of catapults that could hurl rocks and other missiles over varying distances at enemy troops and ships. Finally, there was a collection of mirrors arranged to focus sunlight in such a way as to set enemy ships on fire. This last development is open to a great deal of historic and technical speculation concerning its efficacy. Whether Archimedes successfully used mirrors to set Roman ships on fire during the prolonged siege of Syracuse in the Second Punic War is not known for certain, but his other machines are known to have inflicted a great number of casualties on the attacking Romans.

After Syracuse fell in 212 B.C.E. and the city was being sacked, Archimedes was killed by a Roman soldier. The soldier slew the aging Greek engineer despite standing orders from the Roman general Marcellus that the brilliant man be taken alive and treated with dignity.

The Roman historian Plutarch reported several accounts concerning the death of Archimedes. Two of these accounts are mentioned here. In the first account, Archimedes is murdered by a Roman soldier out of retribution, since the soldier wanted payback for all his comrades who had perished because of Archimedes' machines. The other account suggests that as the city fell, a Roman soldier suddenly came upon Archimedes, who was sitting on the ground drawing circles and other geometric figures in the sand. When told to move, the absentminded Archimedes ignored the order and asked for time to finish the geometry problem. The impatient Roman gave him a fatal thrust with a short sword instead.

## ANCIENT TOYS ANTICIPATE MODERN TURBINES

Heron of Alexandria (ca. 20–80 C.E.) (also known as Hero) was the last of the great Greek engineers of antiquity. He invented many clever mechanical devices, including the aeoliphile, the prophetic toylike device for which he is mostly remembered. The aeoliphile was a spinning, steam-powered spherical apparatus that demonstrated the action-reaction principle, the physical principle that forms the basis of Newton's third law of motion and modern reaction devices.

Not much has survived from antiquity about the personal life of Heron. Historians estimate that he was born in about 20 C.E. because his own writings suggest that he witnessed the lunar eclipse that was observable in Alexandria in 62 C.E. Heron had a strong interest in simple machines, mechanical mechanisms such as gears, and hydraulic and pneumatic systems. His inventions and publications reflect the influence of Ctesibius. Several of Heron's works have survived, including *Pneumatics* (written about 60 C.E.), *Automata, Mechanics, Dioptra,* and *Metrics.*

His most familiar invention is the aeoliphile. He placed a hollow metal sphere on pivots over a heating device that resembled a charcoal grill. When water placed inside the metal sphere was heated over the brazier, steam formed and escaped through tubes, which acted like crude nozzles. The sphere would spin freely as steam escaped from two small opposing tubes connected to the sphere. This whirling sphere delighted children and became a popular toy. However, for some inexplicable reason, Heron never connected the action-reaction principle exhibited by the aeoliphile with the concept of spinning, steam-powered machines for performing useful work.

An artist's drawing of Heron of Alexandria's aeoliphile—a primitive steam engine *(Modeled after NASA artwork)*

The aeoliphile is an example of a clever device invented well ahead of its time. Such devices sometimes need to be rediscovered or reinvented decades or centuries later, when the social, economic, and/or technical conditions are right for full engineering development and application. Since the aeoliphile embodies the action-reaction principle, it is the technical ancestor of the steam turbine, which helped industrialize (and later electrify) the world, as well the powerful rocket vehicles that sent robot spacecraft to explore distant worlds in the solar system.

Despite this oversight, Heron remained a skilled engineer and creative inventor. According to translations of his surviving documents, he designed an automated temple door complex that used an elaborate collection of pulleys, pipes, and buckets of water—all hidden from view of the worshipers. When a sacrificial fire burned on the altar, people in the temple became astonished as a set of special doors would slowly open. Heron used heat from the altar fire to expand the air in the hollow sphere he had hidden below the altar. As the heated air expanded, it slowly expelled water from the sealed sphere through a pipe into a bucket. As the bucket filled with water, it became heavier and descended on a cord. The downward motion of the bucket pulled a cord, which turned two drums, automatically opening the temple doors and raising a counterweight. When the altar fire was extinguished, the air in the hollow sphere cooled. This drop in temperature created a partial vacuum that drew water back into the sphere from the bucket. Finally, as water left the bucket, its weight decreased, and the counterweight began to descend and close the temple doors.

Heron also receives credit for other feedback control devices that used fire, water, and compressed air in different combinations. He developed a machine for threading wooden screws and constructed an automated puppet theater. According to surviving historic accounts, he also designed an early odometer; a primitive form of analog computer that involved gears, spindles, weights, pegs, trays of sand, and ropes; and a compressed air fountain.

# CHAPTER 2

# Physical Characteristics of Liquids

**This chapter describes how scientists use certain macroscopic (bulk)** properties of liquids, such as density, pressure, temperature, and viscosity, to explain and predict their physical behavior. Underlying this important process are many careful measurements performed within a system of scientific units.

## UNDERSTANDING THE PROPERTIES OF MATTER

Scientists often compare physical and chemical properties in their efforts to fully characterize substances. A physical property of a substance is a measurable behavior or physical characteristic. Some of the more common physical properties used by scientists and engineers are mass, density, temperature, structure, and hardness. For liquids, other important physical properties include melting point, boiling point, and viscosity. Heat capacity, thermal conductivity, electric conductivity, and surface tension are still other significant physical properties.

The chemical properties of a substance describe how that particular substance reacts with other substances. Viewed on a microscopic scale, the atoms of the substance experience change while interacting with nearby atoms. Electron trading among neighboring atoms makes or breaks bonds. Changes in ionization state (from electrically neutral to positive or negative) may also take place at the atomic level. As a result of their chemical properties, some substances experience corrosion; others experience combustion (burning) or explosive decomposition; while others remain inert and refuse to interact with different materials in their environment. Today, chemists and physicists correlate the properties of the elements portrayed in the periodic table (see appendix) with their electron configurations.

Scientists regard a physical change of a substance as one that involves a change in the physical appearance of that particular sample of matter, but the matter experiencing the physical *change of state* does not undergo any change in its composition or chemical identity. Consider adding heat to an uncovered pot of water on a stove. The water eventually starts to boil and then experiences a physical change by becoming a hot gas commonly called steam. Scientists say that water experiences a physical change of state when it transforms to steam. Despite this change in physical state, steam retains the chemical identity of water.

Each of the three commonly encountered states of matter (solid, liquid, and gas) can change into either of the other two states by undergoing a change of state (or phase transition). The addition or removal of energy from the substance, usually as heat, facilitates such changes of state. The transitions are sensitive to the nature of the substance as well as its tem-

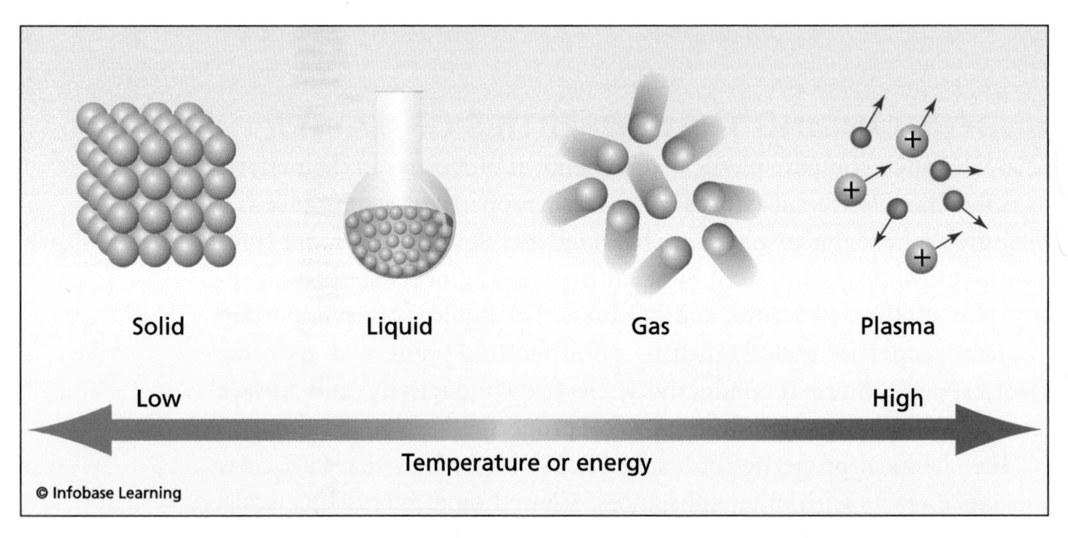


As energy is added to a solid, its temperature increases, and the solid becomes a liquid. Further addition of energy allows the liquid to become a gas. If the gas is heated a great deal more, its atoms break apart into charged particles, resulting in the fourth state of matter, plasma. *(Modeled after NASA–sponsored artwork)*

perature and pressure. Scientists and engineers have developed a collection of technical terms to describe changes of state. They define melting as the change of a substance from the solid state to the liquid state; *freezing* is the opposite process. They define the melting point as the temperature, for a specified pressure, at which a substance transforms from the solid state to the liquid state. At this temperature, the solid and liquid states of a substance can coexist in equilibrium. The term *melting point* is synonymous with the term *freezing point.*

Scientists use the term *vaporization* to describe the process by which a substance in the liquid state becomes a gas. The term *evaporation* is also used. In the reverse process, called condensation, a substance in the gas (vapor) state becomes a liquid. When the evaporation rate of a liquid equals the condensation rate of the vapor, scientists say the liquid and vapor (gas) states of the particular substance are in equilibrium. They define the *boiling point* as the temperature, at a specified pressure, at which a substance in the liquid state experiences a change into the gas state. When a substance such as oxygen or nitrogen that is normally encountered as a gas on Earth experiences a change to the liquid state, scientists call the process liquefaction.

Sometimes a solid substance is at a temperature and pressure that allows it to transition directly into the gas (vapor) state. Scientists call this process sublimation. A frozen block of carbon dioxide ($CO_2$) at room temperature and one atmosphere pressure will transition (sublime) directly into gaseous carbon dioxide. Scientists refer to the reverse process as deposition. Under certain environmental conditions, water vapor in Earth's atmosphere transitions directly into the familiar solid known as snow. Once formed, snowflakes gently deposit themselves on the planet's surface.

Vapor pressure is an important thermodynamic property of a liquid substance. Engineers regard it as a useful measure of a liquid's inclination to evaporate. Volatile substances are those with high vapor pressures at normal temperatures and pressures. Chemists use the term *volatility* to describe the tendency of a liquid to vaporize. Consider a sealed flask of liquid with a sufficient unfilled volume (space) above the liquid's free surface. Under equilibrium conditions, a number of molecules in the liquid will escape across the free surface and occupy the open space as gas molecules. Scientists define the vapor pressure of a liquid as the partial pressure of the vapor over the liquid under equilibrium conditions at a specified temperature. The science of multistate (multiphase) substances is quite complicated. This book primarily focuses on substances in the liquid state.

In general, a solid occupies a specific, fixed volume and retains its shape. A liquid also occupies a specific volume but is free to flow and assume the shape of the portion of the container it occupies. A gas has neither a definite shape nor a specific volume. Rather it will quickly fill the entire volume of a closed container. Unlike solids and liquids, which resist compression (or squeezing), gases are compressed easily. The term *fluid* includes both liquids and gases. On Earth, water is the most commonly encountered liquid, and air is the most commonly experienced gas.

During the Scientific Revolution, pioneering scientists such as Galileo Galilei, Sir Isaac Newton, Evangelista Torricelli, Blaise Pascal, and Daniel Bernoulli began describing and predicting fluid behavior. Their technical efforts yielded important fluid science relationships for both liquids and gases. They employed interesting experiments and mathematical relationships to help unlock nature's secrets. Their pioneering activities formed the broad field of fluid science, also termed fluid mechanics.

Fluid science is the major branch of science that deals with the behavior of fluids (both gases and liquids) at rest (fluid statics) and in motion (fluid dynamics). This scientific field has many subordinate branches and important applications, including aerodynamics (the motion of gases, including and especially air), hydrostatics (liquids, including water) at rest, and hydrodynamics (liquids in motion, including naturally flowing and artificially pumped water). This book concentrates on the interesting phenomena and scientific principles associated with liquids and gives special attention to the most important liquids in the daily lives of most people—water, petroleum (oil), and bodily fluids (such as blood and urine).

A fluid is a substance that, when in static equilibrium, cannot sustain a shear stress. A perfect (or ideal) fluid is one that has zero viscosity—that is, offers no resistance to shape change. Actual fluids only approximate this behavior. Viscosity is an important idea linked to internal fluid friction. This phenomenon is quite apparent when handling liquids. For example, water is much less viscous than honey. Even though both are liquids at room temperature, because of differences in their viscosity values, they flow at significantly different rates.

When scientists and engineers use the term *fluid,* they mean both liquids and gases. One easy way to understand the difference between the two is that a liquid will take the shape of and stay in an uncovered container, such as a bowl or coffee cup, while a gas also takes the shape of its

Honey is an example of a viscous fluid. Many people enjoy drizzling this naturally sweet, organic liquid over freshly baked biscuits. *(USDA/ Agricultural Research Service)*

container but will escape unless the container is tightly sealed. Scientists call the open surface of a liquid in a container the free surface. The surface of the ocean is a free surface that defines the important natural boundary between the hydrosphere and the atmosphere. Evaporation and condensation take place at the free surface as part of Earth's great water cycle.

## SCIENTIFIC UNITS

Science is based upon repeatable experiments, accurate measurements, and a system of logical, standardized units. The International System of Units is the generally agreed upon coherent unit system now in use throughout the world for scientific, engineering, and commercial purposes. Universally abbreviated SI (from the French Système International d'Unités), people also call it the metric system.

The contemporary SI system traces its origins to France in the late 18th century. During the French Revolution, government officials decided to introduce a decimal measurement system. As a complement to the physical standards for length and mass carefully fabricated out of platinum, scientists used astronomical techniques to precisely define a second of time.

Scientists based the fundamental SI units for length, mass, and time—the meter (m) (British spelling: *metre*), the kilogram (kg), and the second (s)—on natural standards. Modern SI units still rely on natural standards and international agreement, but the new standards can be measured with much greater precision. Originally, scientists defined the meter as one ten-millionth the distance from the equator to the North Pole along the meridian of longitude nearest Paris. They denoted this standard of length with a carefully protected platinum rod. In 1983, scientists refined the SI definition of length with the following statement: "The meter is the length of the path traveled by light in vacuum during a time interval of 1/299,792,458 of a second." Scientists now use the speed of light in a vacuum (namely 299,792,458 m/s) as the natural standard for defining the meter.

Other basic units in the SI system are the ampere (A) (a measure of electric current), the candela (cd) (a measure of luminous intensity), the kelvin (K) (a measure of absolute thermodynamic temperature), and the mole (mol) (a measure of the amount of substance). The radian (rad) is the basic SI unit of plane angle and the steradian (sr) the basic SI unit of solid angle. There are also numerous supplementary and derived SI units, such as the newton (N) (a measure of force) and the pascal (Pa) (a measure of pressure).

Today, SI measurements play an important role in science and technology throughout the world. However, considerable technical and commercial activity in the United States still involves the use of another set of units, known as the American customary system of units. This system traces its heritage back to colonial America and the weights and measures then being used throughout the 18th-century British Empire.

One major advantage of SI units over American customary units is that SI units employ the number 10 as a base. Multiples or submultiples of SI units are easily reached by multiplying or dividing by 10. The inch (in) and the foot (ft) are the two commonly encountered units of length in the American customary system. Both originated in medieval England and were based upon human anatomy. The width of a human thumb gave rise to the inch; the human foot, the foot. Through an evolving process of standardization, 12 inches became equal to one foot; three feet equal to one yard (yd).

The SI system uses the kilogram as the basic unit of mass, while the American customary system uses the pound-mass (lbm). The American system also contains another fundamental "pound" unit, the pound-force (lbf), which is actually a unit of force. Both "pounds" were related by an arbitrary decision made centuries ago to create a measurement system consisting of four basic (fundamental) units: length, mass, time, and force. To implement this decision, officials declared that "a one pound-mass object exerts (weighs) one pound-force at sea level." Technical novices and professionals alike all too often forget that this arbitrary pound equivalency is valid only at sea level on the surface of Earth. The historic arrangement that established the pound-force as a basic (or fundamental) unit rather than a derived unit (based on Newton's second law of motion) persists in the American customary system and can cause considerable confusion.

Scientists use SI units in their work to avoid ambiguity and potential confusion. Many people find it helpful to remember that the comparable SI unit for the pound-mass is the kilogram (kg), namely 1 kg = 2.205 lbm, while the comparable SI unit for the pound-force is called the newton (N), namely 1 N = 0.2248 lbf.

## FORCE, ENERGY, AND WORK

The science of mechanics links the motion of a material object with the measurable quantities of mass, velocity, and acceleration. Through Newton's efforts, the term *force* entered the lexicon of physics. Scientists say a force influences a material object by causing a change in its state of motion. (Historically, Newton used the term *fluction* when he developed the calculus.) The concept of force emerges out of Newton's second law of motion. In its simplest and most familiar mathematical form, force *(F)* is the product of an object's mass (m) and its acceleration *(a)*—namely, $F = m\ a$. In his honor, scientists call the fundamental unit of force in the SI system the newton (N). A force of one newton accelerates a mass of one kilogram at the rate of one meter per second per second ($1\ N = 1\ kg\text{–}m/s^2$). In the American customary system, engineers define one pound-force (lbf) as the force equivalent to the weight of a one-pound mass (1 lbm) object at sea level on Earth.

Physicists define *energy* (E) as the ability to do work. Within classical Newtonian physics, scientists describe mechanical work (W) as a force *(F)* acting through a distance (d). The amount of work done is proportional to both the force involved and the distance over which the force is exerted in the direction of motion. A force perpendicular to the direction of motion performs no work. In the SI system, scientists measure energy with a unit

## NEWTON'S LAWS OF MOTION

Newton's laws of motion are the three fundamental postulates that form the basis of classical mechanics. He formulated these laws in about 1685 while studying the motion of the planets around the Sun. In 1687, Newton presented his work to the scientific community in *The Principia.*

Newton's first law of motion is concerned with the principle of inertia. It states that if a body in motion is not acted upon by an external force, the momentum remains constant. Scientists also refer to this law as the law of the conservation of momentum.

Newton's second law states that the rate of change of momentum of a body is proportional to the force acting upon the body and is in the direction of the applied force. A familiar statement of this law is the equation $\boldsymbol{F} = m\ \boldsymbol{a}$, where $\boldsymbol{F}$ is the vector sum of the applied forces, m is the mass, and $\boldsymbol{a}$ represents the acceleration vector of the body.

Newton's third law is the principle of action and reaction. It states that for every force that acts upon a body, there is a corresponding force of the same magnitude exerted by the body in the opposite direction.

called the joule (J). One joule represents a force of one newton (N) moving through a distance of one meter (m). This unit honors the British physicist James Prescott Joule. In the American customary system, engineers often express energy in terms of the British thermal unit (Btu). (One Btu equals 1,055 joules.)

In classical physics, energy (E), work (W), and distance (d) are scalar quantities, while velocity *(v)*, acceleration *(a)*, and force *(F)* are vector quantities. A scalar is a physical quantity that has magnitude only, while a vector is a physical quantity that has both magnitude and direction at each point in space.

One of the most important contributions of Western civilization to the human race was the development the scientific method and, through it, the start of all modern science. During the intellectually turbulent 17th century in western Europe, people of great genius began identifying important physical laws and demonstrating experimental techniques that helped humans everywhere better understand the physical universe.

## CONCEPTS OF DENSITY, PRESSURE, AND TEMPERATURE

This section discusses three familiar physical properties, density, pressure, and temperature. By thinking about the atoms that make up different materials, scientists are now able to understand, quantify, and predict how interplay at the atomic (microscopic) level results in the physical properties that are measurable on a macroscopic scale.

### Density

To assist in more easily identifying and characterizing different materials, scientists devised the material property called density, one of the most useful macroscopic physical properties of matter. Solid matter is generally denser than liquid matter, and liquid matter denser than gases. Scientists define *density* as the amount of mass contained in a given volume. They frequently use the lower-case Greek letter rho $(\rho)$, as the symbol for density in technical publications and equations.

Scientists use the density $(\rho)$ of a material to determine how massive a given volume of that particular material is. Furthermore, equal volumes of different materials usually have different density values. Density is a function of both the atoms from which a material is composed as well as how closely packed the atoms are arranged in the particular material.

At room temperature (nominally 68°F [20°C]) and one atmosphere pressure, the density of some familiar materials is as follows: gold 1,205 lbm/ft$^3$ (19,300 kg/m$^3$ [19.3 g/cm$^3$]); iron 493 lbm/ft$^3$ (7,900 kg/m$^3$ [7.9 g/cm$^3$]); diamond (carbon) 219 lbm/ft$^3$ (3,500 kg/m$^3$ [3.5 g/cm$^3$]); aluminum 169 lbm/ft$^3$ (2,700 kg/m$^3$ [2.7 g/cm$^3$]); bone 112 lbm/ft$^3$ (1,800 kg/m$^3$ [1.8 g/cm$^3$]); and water 62.3 lbm/ft$^3$ (997.4 kg/m$^3$)[0.997 g/cm$^3$]. Like most gases at room temperature and one atmosphere pressure, oxygen has a density of just 0.083 lbm/ft$^3$ (1.33 kg/m$^3$ [$1.33 \times 10^{-3}$ g/cm$^3$])—a value that is about 1,000 times lower than the density of most solid or liquid materials normally encountered on Earth's surface.

The maximum density of water (pure) occurs when the liquid's temperature is at 39.16°F (3.98°C). Upon freezing, water transforms to ice, which has a lower value of density, namely, 56.81 lbm/ft$^3$ (910 kg/m$^3$ [0.91 g/cm$^3$]) at 32°F (0°C). This decrease in density upon freezing is the reason why ice floats on water—a very important yet unusual natural phenomenon. Most materials have a higher density when solid than when liquid.

Scientists know that the physical properties of matter are often interrelated. Namely, when one physical property (such as temperature) increases or decreases, other physical properties (such as volume or density) also

## SPECIFIC GRAVITY

In many fluid flow problems, engineers find it necessary to include the density of the fluid. To assist in their computations, they define a dimensionless property termed the *specific gravity.* The specific gravity of a substance is the density of that substance divided by the density of a standard reference material. For liquids, the usual reference density is that of pure water at 39.2°F (4°C) and one atmosphere pressure. For this reference density, the specific gravity values of some common liquids are: 0.66 for gasoline (at 32°F [0°C]); 0.807 for ethyl alcohol (at 32°F [0°C]); 0.899 for benzene (at 32°F [0°C]); 1.260 for glycerin (at 32°F [0°C]); and 13.546 for mercury (at 68°F [20°C]). Engineers use a device called a hydrometer to measure the specific gravity of an unknown liquid. The specific gravity of urine plays an important role in medical diagnoses. In comparing gases, engineers often use the density of dry air at a temperature of 68°F (20°C) and one atmosphere pressure as the reference standard. Scientists prefer to use the term *relative density* instead of specific gravity.

## POROSITY AND PERMEABILITY

Scientists define *porosity* as the ratio of pore (void) volume to total volume of a granular substance. The more porous a substance is, the more voids, or empty spaces, per unit volume it has. Beach sand and sandy soils have large voids (pores) and, therefore, higher values of porosity than do clays and other fine-grained soils. Scientists define *permeability* as the ability of a porous solid to allow the passage of a fluid through it when the flowing fluid experiences a (hydraulic) pressure differential. An impermeable material has poorly connected voids and so greatly restricts or totally prohibits the passage of a fluid. When the voids of a porous rock or granular material are well interconnected, a fluid can flow through the substance with relative ease. In sandstone, for example, the well-rounded sand grains create an ample number of interconnected void spaces, so sandstone represents a high porosity and a high permeability solid material.

change. As a direct result of the Scientific Revolution, people learned how to define the behavior of materials by developing special mathematical expressions termed *equations of state*. Scientists created these mathematical relationships using both theory and empirical data from many carefully conducted laboratory experiments.

## Pressure

Scientists describe pressure (P) as force per unit area. The most commonly encountered American unit of pressure is pounds-force per square inch (psi). In the SI system, the fundamental unit of pressure is termed the pascal (Pa) in honor of Blaise Pascal (1623–62). The 17th-century French scientist conducted many pioneering experiments in fluid mechanics. One pascal represents a force of one newton (N) exerted over an area of one square meter—that is, $1 \text{ Pa} = 1 \text{ N/m}^2$. One psi is approximately equal to 6,895 Pa.

Anyone who has plunged into the deep end of a large swimming pool and then descended to the bottom of the pool has personally experienced the phenomenon of hydrostatic pressure. Hydrostatic pressure is the pressure at a given depth below the surface of a static (nonmoving) fluid. As Pascal observed in the 17th century, the greater the depth, the greater the pressure.

## SUPERCRITICAL FLUID

Scientists define the critical pressure of a substance as the highest pressure under which the liquid and gaseous states of a substance can coexist. Similarly, the critical temperature is the temperature above which a substance cannot exist in the liquid state, regardless of the pressure. Above the critical temperature, the liquid smoothly transforms to the gaseous (vapor) state, and boiling disappears. The critical point occurs at the highest temperature and pressure at which liquid and gaseous (vapor) states of a substance can coexist. Above the critical point, scientists call the fluid a supercritical fluid.

Within the supercritical fluid area shown in the accompanying diagram, the fluid exists in only one physical state and possesses both liquidlike and gaslike properties. Supercritical fluids have interesting thermodynamic properties. The physical characteristics of a supercritical fluid resemble those of dense gas, with solubilities approaching values experienced in the liquid state and dynamic viscosities intermediate between the values found in the liquid and gaseous states of the substance. The supercritical properties of carbon dioxide ($CO_2$), for example, make the abundant, nontoxic, and nonflammable substance attractive in industrial applications that require inexpensive solvents.

**_(opposite)_ This plot describes the relationship between pressure and temperature for a typical substance in the liquid and gas state. Beyond the critical point, the substance becomes a supercritical fluid that exhibits both gaslike and liquidlike properties. _(Modeled after DOE/Pacific Northwest National Laboratory)_**

Atmospheric pressure plays an important role in many scientific and engineering disciplines. In an effort to standardize their research activities, scientists use the following equivalent atmospheric pressure values for sea level: one atmosphere (1 atm) ≡ 760 mm of mercury (Hg) (exactly) = 29.92 in (Hg) = 14.695 psi = $1.01325 \times 10^5$ Pa.

One important feature of Earth's atmosphere is that the density in a column of air above a point on the planet's surface is not constant. Density and atmospheric pressure decrease with increasing altitude, until both become negligible in outer space. The pioneering work of Pascal and the Italian physicist Evangelista Torricelli (1608–47) guided other scientists in measuring and characterizing Earth's atmosphere.

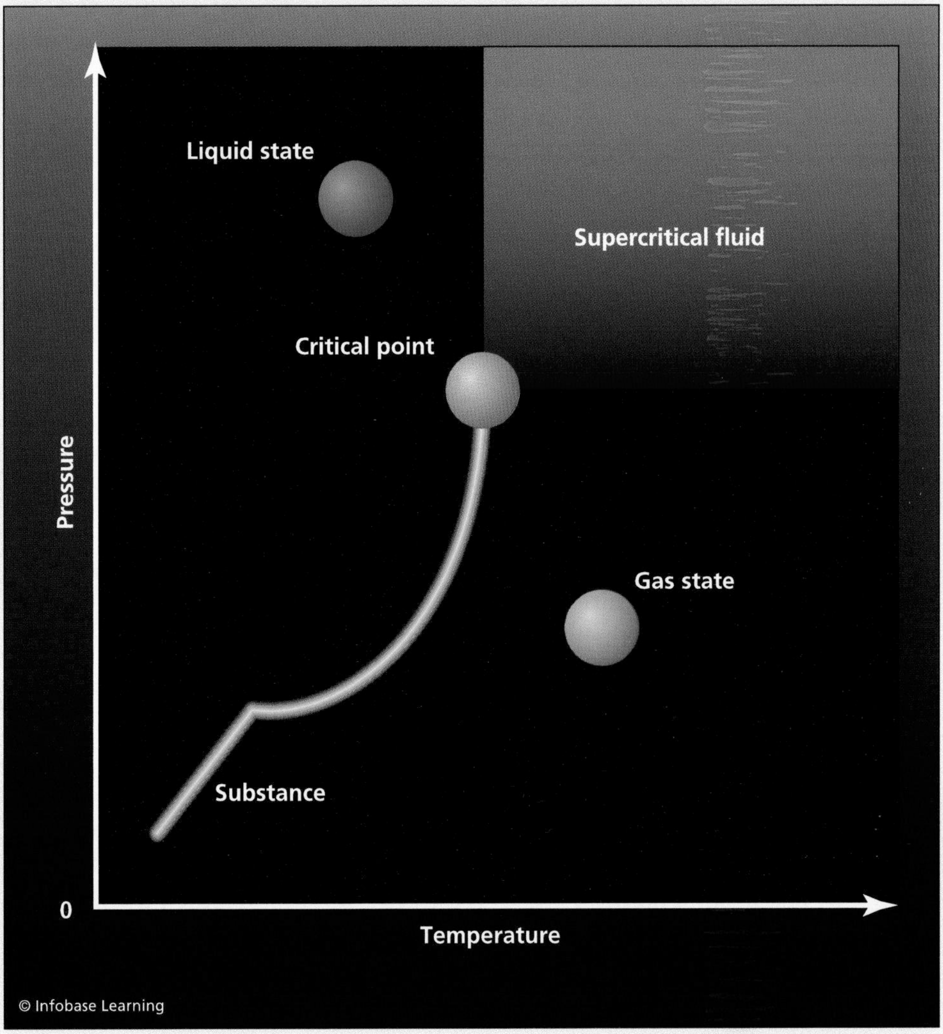

Engineers often treat rigid, solid bodies as incompressible objects. Generally, very high values of force are needed to compress or deform a rigid, solid body. Unlike rigid solids, fluids are materials that can flow, so engineers use pressure differentials to move fluids. They design pumps to move liquids (often treated as incompressible fluids), while they design fans to move gases (compressible fluids). An incompressible fluid is assumed to have a constant value of density; a compressible fluid has a variable density. One of the interesting characteristics of gases is that, unlike solids or liquids, they can be compressed into smaller and smaller volumes.

## Temperature

While temperature is one of the more familiar physical variables, it is also one of the more difficult to quantify. Scientists suggest that on the macroscopic scale, temperature is the physical quantity that indicates how hot or cold an object is relative to an agreed upon standard value. Temperature defines the natural direction in which energy will flow as heat—namely, from a higher-temperature (hot) region to a lower-temperature (cold) region. Taking a microscopic perspective, temperature indicates the speed at which the atoms and molecules of a substance are moving.

Scientists recognize that every object has the physical property called temperature. They further understand that when two bodies are in thermal equilibrium, their temperatures are equal. A thermometer is an instrument that measures temperatures relative to some reference value. As part of the Scientific Revolution, creative individuals began using a variety of physical principles, natural references, and scales in their attempts to quantify the property of temperature.

In about 1592, Galileo attempted to measure temperature with a device he called the thermoscope. (Scientists customarily refer to Galileo Galilei by just his first name.) Although Galileo's work represented the first serious attempt to harness the notion of temperature as a useful scientific property, his thermoscope, while innovative, did not supply scientifically significant temperature data.

The German physicist Daniel Gabriel Fahrenheit (1686–1736) was the first person to develop a thermometer capable of making accurate, reproducible measurements of temperature. In 1709, he observed that alcohol expanded when heated and constructed the first closed-bulb glass thermometer with alcohol as the temperature-sensitive working fluid. Five years later, in 1714, he used mercury as the thermometer's working fluid. Fahrenheit selected an interesting three-point temperature reference scale for his original thermometers. His zero point (0°F) was the lowest temperature he could achieve with a chilling mixture of ice, water, and ammonium chloride ($NH_4Cl$). Fahrenheit then used a mixture of just water and ice as his second reference temperature (32°F). Finally, he chose his own body temperature (recorded as 96°F) as the scale's third reference temperature.

After his death, other scientists revised and refined the original Fahrenheit temperature scale, making sure there were 180 degrees between the freezing point of water (32°F) and the boiling point of water (212°F) at one atmosphere pressure. On this refined scale, the average temperature of the human body appeared as 98.6°F. Although the Fahrenheit temperature scale is still used in the United States, most other nations have adopted another relative temperature scale called the Celsius scale.

In 1742, the Swedish astronomer Anders Celsius (1701–44) introduced the relative temperature scale that is now carries his name. He initially selected the upper (100-degree) reference temperature on his new scale as the freezing point of water and the lower (0-degree) reference temperature as the boiling of water at one atmosphere pressure. He then divided the scale into 100 units. After Celsius's death, the Swedish botanist and zoologist Carl Linnaeus (1707–78) introduced the present-day Celsius scale thermometer by reversing the reference temperatures. The modern Celsius temperature scale is a relative temperature scale in which the range between two reference points (the freezing point of water at 0°C and the

## RANKINE—THE OTHER ABSOLUTE TEMPERATURE

Most of the world's scientists and engineers use the Kelvin scale to express absolute thermodynamic temperatures, but there is another absolute temperature scale, the Rankine scale (symbol R), that sometimes appears in engineering analyses performed in the United States—analyses based upon the American customary system of units. In 1859, the Scottish engineer William John Macquorn Rankine (1820–72) introduced the absolute temperature scale that now carries his name. Absolute zero in the Rankine temperature scale (that is, 0 R) corresponds to −459.67°F. The relationship between temperatures expressed in rankines (R) and those expressed in the degrees Fahrenheit (°F) is T (R) = T (°F) + 459.67. The relationship between the Kelvin scale and the Rankine scale is (9/5) × absolute temperature (kelvins) = absolute temperature (rankines). For example, a temperature of 100 K is expressed as 180 R. The use of absolute temperatures is very important in science.

boiling point of water at 100°C) are conveniently divided into 100 equal units, or degrees.

Scientists describe a relative temperature scale as one that measures how far above or below a certain temperature measurement is with respect to a specific reference point. The individual degrees, or units, in a relative scale are determined by dividing the relative scale between two known reference temperature points (such as the freezing and boiling points of water at one atmosphere pressure) into a convenient number of temperature units (such as 100 or 180).

Despite considerable progress in thermometry in the 18th century, scientists still needed a more comprehensive temperature scale—namely, one that included the concept of absolute zero, the lowest possible temperature, at which molecular motion ceases. The Irish-born British physicist William Thomson Kelvin (1824–1907), first baron of Largs, proposed an absolute temperature scale in 1848. The scientific community quickly embraced Kelvin's scale. The proper SI term for temperature is *kelvins* (without the word *degree*), and the proper symbol is K (without the symbol °). Scientists generally use absolute temperatures in such disciplines as physics, astronomy, and chemistry; engineers use either relative or absolute temperatures in thermodynamics, heat transfer analyses, and mechanics, depending upon the nature of the problem. Absolute temperature values are always positive, but relative temperatures can have positive or negative values.

## VISCOSITY

Scientists use the property of viscosity as a measure of the internal friction, or flow resistance, of a fluid when the fluid is subjected to shear stress ($\tau$). In the 17th century, Newton conducted relatively simple flow experiments that allowed him to make two important conclusions about fluid friction. His research still supports a basic understanding of the flow mechanics of real fluids.

The accompanying illustration depicts viscous flow between parallel plates (walls) in relative motion. The lower plate is assumed fixed in place, while the upper plate is moving. Newton's first important observation was that the fluid does not move (or slide) along the fixed wall. This conclusion represents the important "no-slip" hypothesis in fluid mechanics. Since the fluid adheres to the surface of the stationary (fixed) wall, the fluid velocity (V) is zero at this boundary. (For simplicity, vector notation is omitted here.) When scientists and engineers analyze frictional flow resistance in contemporary problems, they usually start by assuming that relative flow velocity at every point on the surface of a solid fixed boundary is zero. The German physicist Ludwig Prandtl (1875–1953) expanded Newton's work by introducing boundary layer theory in the early 20th century.

Newton's second conclusion involved his recognition that the resistive (viscous) force (F) on the moving upper wall is directly proportional to the relative velocity (V) and inversely proportional to the distance (y) separating the parallel walls. Assuming the shear stress ($\tau$) equals the force (F) divided by the area (A), Newton and subsequent scientists were able to write $\tau = F/A = \mu\ (V/H)$, where V is the velocity of the moving plate at the distance ($y = H$) from the fixed wall and $\mu$ is the dynamic viscosity coefficient. In more elaborate mathematical treatments of boundary layer phenomenon, differential calculus plays an important role. It allows scientists to express the shear stress for a fluid element at any point in the flow as follows: $\tau = \mu\ (\partial V/\partial y)$. The ratio $\partial V/\partial y$ represents the velocity

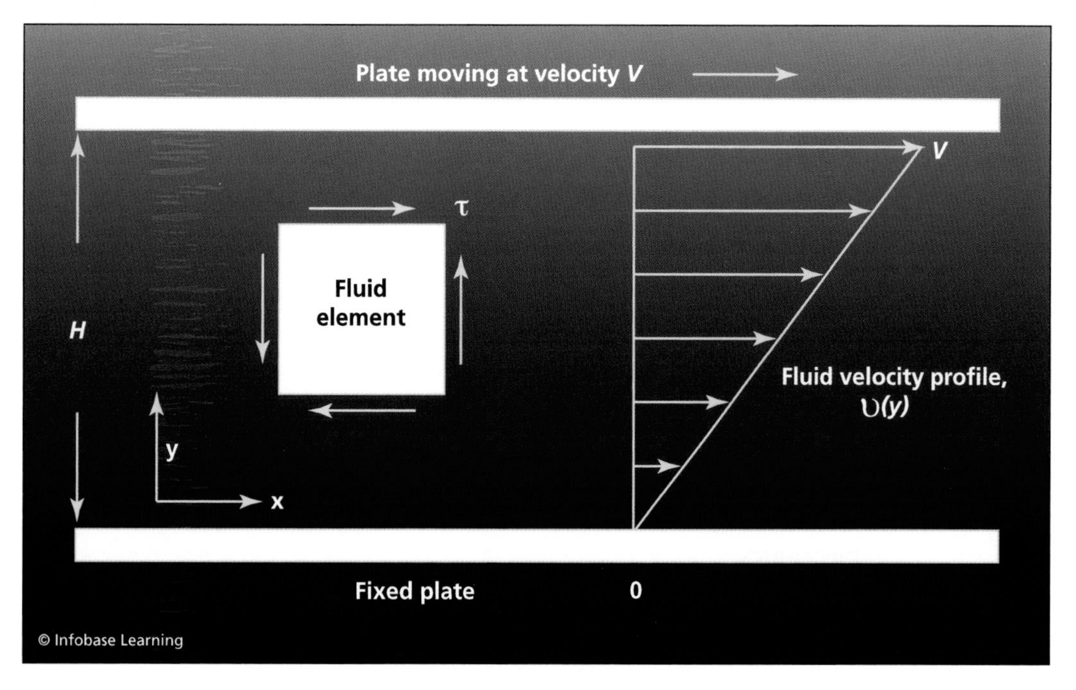


Diagram depicting Newton's classic viscosity experiment involving two parallel plates (walls) in relative motion *(Author, modeled after NASA artwork)*

gradient, or rate of shearing strain, and μ the constant of proportionality. Scientists refer to the term $\mu$ as the coefficient of viscosity, or the dynamic viscosity. In American customary units, scientists quantify μ as (pounds-force-second)/square foot, or ($lbf\text{-}s/ft^2$); in SI units, μ is expressed as (newtons-second)/square meter, or ($N\text{-}s/m^2$). Since scientists define force per unit area as pressure, the dynamic viscosity becomes pascal–seconds (Pa-s) in SI units. Physically, the dynamic viscosity represents the force that must be applied per unit area to permit adjacent layers of fluid to move with unit velocity relative to each other.

Scientists define kinetic viscosity (ν) as the dynamic viscosity divided by the fluid's density. In SI units, scientists express kinematic viscosity as $m^2/s$; in American customary units, as $ft^2/s$. At room temperature (nominally 68°F [20°C]) and one atmosphere pressure, water has a kinematic viscosity of approximately $1.076 \times 10^{-5}$ $ft^2/s$ ($1.0 \times 10^{-6}$ $m^2/s$); unused engine oil a kinematic viscosity of 0.00969 $ft^2/s$ (0.00090 $m^2/s$); glycerin ($C_3H_5[OH]_3$) a kinematic viscosity of 0.0127 $ft^2/s$ (0.00118 $m^2/s$); and mercury (Hg) a kinematic viscosity of $1.227 \times 10^{-6}$ $ft^2/s$ ($0.114 \times 10^{-6}$ $m^2/s$). Glycerin is a colorless, viscous liquid that scientists use in pharmaceutical applications. In general, the viscosity of a liquid decreases with increasing fluid temperature; for a gas, the viscosity increases with increasing fluid temperature.

Fluids for which the shear stress (τ) is proportional to the velocity gradient ($\partial V/\partial y$) are called *Newtonian fluids*. Scientists treat most gases and common liquids such as water, oil, and honey as Newtonian fluids. Other fluids such as mayonnaise, peanut butter, paint, and molten plastics have more complex viscous flow behavior, so scientists treat them as non-Newtonian fluids. Although all real fluids (except superfluids) exhibit some internal resistance to flow, scientists sometimes find it convenient as a first approximation to treat the fluid as a perfect (or ideal) fluid that experiences no (or at least negligible) resistance to flow. They call the perfect fluid with zero viscosity an inviscid fluid. When some degree of viscosity influences flow, scientists term the fluid a *viscid fluid*. The study of how viscosity influences flow is a fascinating and complex part of fluid mechanics.

# CHAPTER 3

# Fundamental Principles of Fluid Science

**This chapter introduces some of the basic scientific principles that** govern the science of fluids. Three key ideas establish the overall architecture of fluid science. These foundational concepts are Newton's laws of motion, the continuity principle (indestructibility of flowing matter), and the conservation of energy. Scientists also use several other important concepts in their investigation of fluids. The additional concepts include the idea of pressure (based upon Pascal's principle of hydrostatics), the concept of buoyancy (based upon the Archimedes principle), the perfect gas law (based upon the work of Robert Boyle [1627–91] and other pneumatic chemists), and the Bernoulli principle. Emphasis is placed on the behavior of liquids, especially water, both at rest (hydrostatics) and in motion (hydrodynamics).

As a starting point in their analytical work, scientists and engineers usually assume that liquids are incompressible fluids. For many common liquids, this assumption is a remarkably good first approximation. When more sophisticated analyses are needed, scientists and engineers use computational fluid dynamics (CFD).

## FLUID STATICS

The field of fluid statics (hydrostatics) addresses the behavior of fluids at rest. Pascal observed that the pressure of a fluid at rest is the same in all directions. His work introduced the notion of hydrostatic pressure and represents the foundational principle within fluid statics. The concept of buoyancy, as embraced within the Archimedes principle, represents another important scientific concept within fluid statics. This section describes how scientists use some of these basic concepts to understand and predict the behavior of liquids.

## Pascal's Principle

In the 17th century, the French scientist Blaise Pascal (1623–62) investigated the properties of liquids at rest. He observed that the pressure in a liquid increased with depth due to the weight of the fluid above the point (depth) of interest. One of the main scientific results of his efforts was the important principle of hydrostatics, now known as Pascal's principle. This principle states that any change in the pressure applied to a completely enclosed liquid is transmitted undiminished to all parts of the liquid and the enclosing container's walls. It is the basic principle governing the operation of an enormous number of devices that use hydraulic devices, including car lifts, hydraulic presses and elevators, syringes, and similar fluid mechanics devices. Pascal recognized that pressure is the physical property that two systems have in common when they are in mechanical equilibrium. Today, scientists more rigorously define pressure as the normal component of force per unit area exerted by a fluid on a boundary.

In the 1650s, Pascal discovered that the pressure at any point in a liquid at rest is the same in all directions. While considering a tall column of liquid, he developed the basic principle of hydrostatics—namely, that the pressure of the liquid increases as the depth increases. Scientists use mathematics to express this important physical principle, as follows: $P = \rho gh$. This simple equation states that the hydrostatic pressure (P) of a column of liquid of height (h) is equal to the product of the liquid's density ($\rho$) (assumed constant), the height (h), and the acceleration of gravity *(g)*. For a given liquid, the hydrostatic pressure depends on the height of the column or depth below some datum, such as sea level. The pressure at any point in the fluid at a specified depth is the same in all directions, because at the microscopic level the fluid pressure is related to the collision of atoms or molecules as they move about in all directions.

Civil engineers who design dams to impound large quantities water must pay careful attention to the hydrostatic pressure on the dam, especially the maximum pressure at the base. (See chapter 6.) Nautical engineers who design submersible vessels must plan for the hydrostatic

A group of visitors enjoying a face-to-face encounter with a Pacific white-sided dolphin at the John G. Shedd Aquarium in Chicago *(© Shedd Aquarium/Brenna Hernandez used with permission)*

pressure at the submarine's maximum operating depth. A submarine that descended to depths significantly below the safe operational limit would experience excessive hydrostatic pressures, resulting in hull crushing and vessel destruction. Similarly, industrial engineers who design large tanks to hold various liquids must design the tank walls and supporting structure to accommodate the hydrostatic pressure within a full tank of liquid.

Many people enjoy visiting large modern aquaria that possess multi-million-gallon (liter) quantities of sea water. Engineers design such facilities with expansive glass windows so visitors can view the marine life within. The glass (or special plastic) used to permit panoramic viewing in these facilities must be strong enough to withstand the hydrostatic pressure that occurs at the bottom of these very large tanks.

## Hydraulic Lever

Engineers use Pascal's principle as the basis for an important machine called the hydraulic lever. The device contains an incompressible fluid that transmits and amplifies an applied force. Since the enclosed fluid is assumed to be incompressible, the pressure ($p_1$) within the liquid at piston one is equal to the pressure ($p_2$) within the liquid at piston two. The work

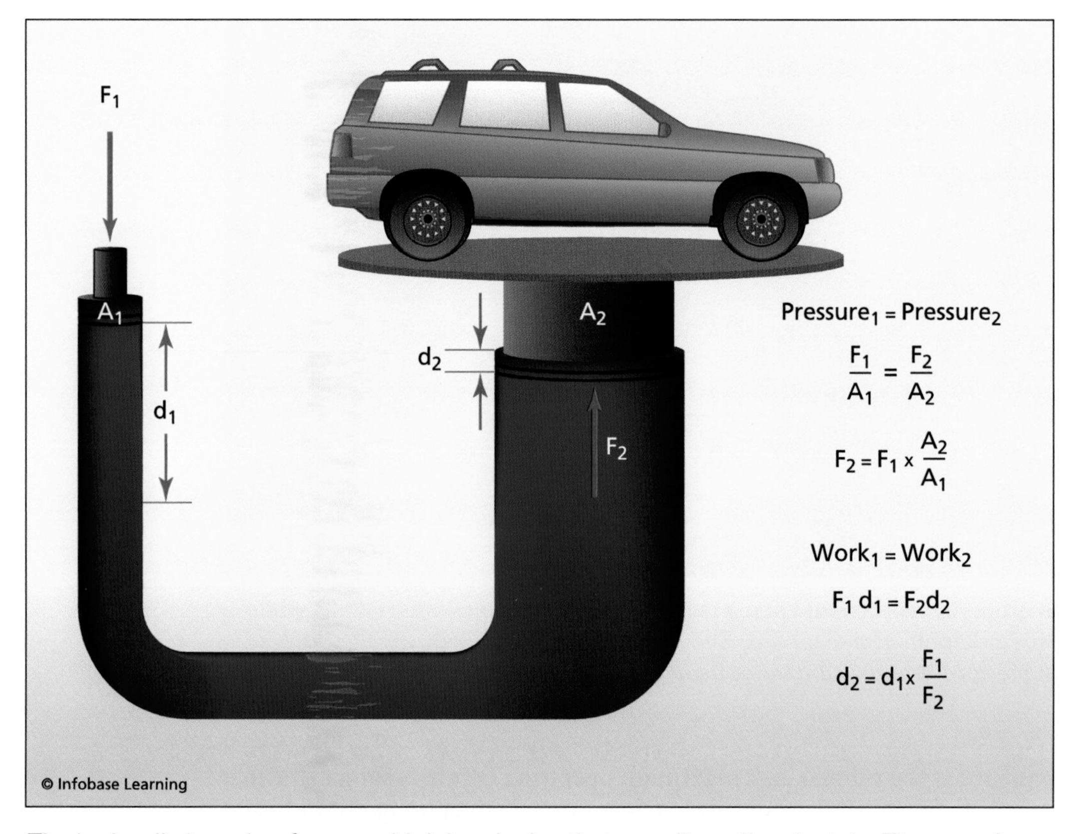


The hydraulic lever is a force-multiplying device that uses Pascal's principle. The use of an incompressible fluid allows a small force ($F_1$) exerted on a small area ($A_1$) applied through a large distance ($d_1$) to become a larger force ($F_2$) exerted over a larger area ($A_2$) applied through a smaller distance ($d_2$).

applied by the person operating the hydraulic device is not magnified. Physicists define work as a force applied through a distance. Neglecting internal friction within the device, the work is the same for both the input and output forces—that is, $work_1 = work_2$. A person applies an input force ($F_1$) to the hydraulic lever at location one, which has an area $A_1$. He or she has to displace (pump) the handle of the hydraulic lever a much farther distance ($d_1$), usually in a series of small strokes, than the distance ($d_2$) to make the massive automobile rise.

What happens through the process of force (not work) amplification is that the force applied on the automobile ($F_2$) is amplified by the hydraulic lever in a manner described mathematically as $F_2 = F_1 (A_2/A_1)$. With the hydraulic lever, a given smaller input force ($F_1$) applied over a larger distance

($d_1$) gets transformed into a greater force ($F_2$) applied over a smaller distance ($d_2$). Although the input work ($W_1$) remains equal to the output work ($W_2$) (used to raise the automobile), the hydraulic lever provides a mechanical advantage by amplifying the magnitude of the force. When a person uses the hydraulic lever, a small force applied to a small area (over a longer distance [$d_1$]) becomes a large force applied to a large area (over a smaller distance [$d_2$]). Physicists mathematically express this circumstance as

$$d_2 = (F_1/F_2)\, d_1.$$

## Archimedes Principle

Another fascinating property of fluids involves the force exerted by a fluid (liquid or gas) on a solid object immersed in the fluid. As the ancient Greek engineer Archimedes (ca. 287–212 B.C.E.) discovered more than two millennia ago, the upward force exerted on an object immersed in a fluid (in this chapter the fluid is assumed to be a liquid) is equal to the weight of the fluid that the object displaces. Physicists define the force exerted by the fluid on the object as the buoyant force, and the phenomenon is called buoyancy. They express the weight of the displaced fluid as its mass times the acceleration of gravity, that is, $m_{fluid} \times g$. Ships made of steel with proper hull designs, hot air balloons, and icebergs all float because of buoyancy. A large supertanker loaded with crude oil can float and successfully travel across the ocean because of the simple fact that the weight of the ship's steel plus the weight of all its cargo, equipment, and crew remains less than the weight of the volume of water displaced by the hull.

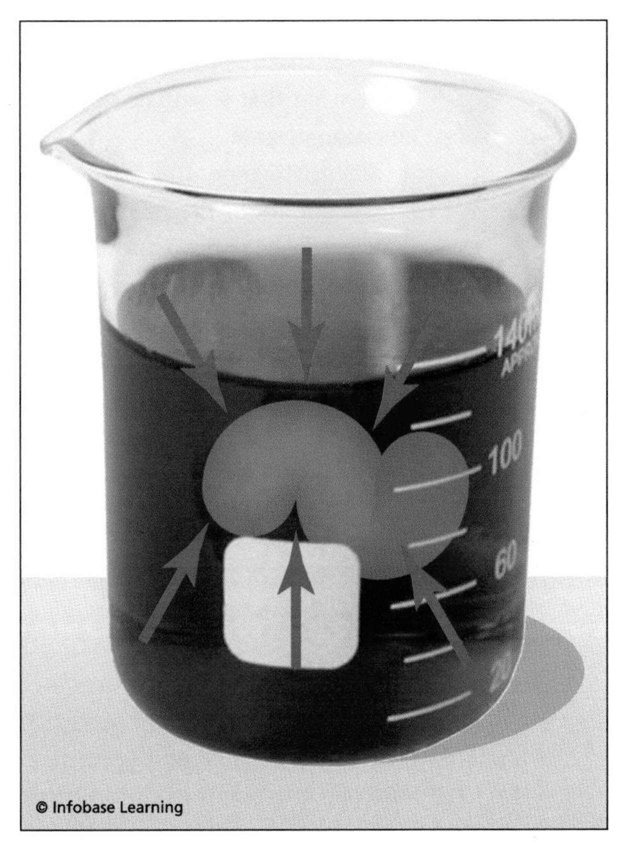



Buoyancy is a manifestation of Archimedes' principle. Scientists say that a fluid exerts a net upward (buoyant) force on a fully or partially submerged object. The buoyant force is directed upward and has a magnitude equal to the weight of the fluid displaced by the object.

## FLUID DYNAMICS

The field of fluid dynamics addresses the behavior of fluids in motion. The motion of real fluids is very complicated and not yet completely understood by scientists and engineers. To make useful predictions about fluid behavior, they need to make simplifying assumptions that make the mathematics and theories a bit more manageable. Scientists usually start their analysis of fluid flow by assuming the fluid is a perfect (ideal) fluid. A perfect fluid is a hypothesized fluid primarily characterized by a lack of viscosity and, usually, by incompressibility. Liquids, such as water, mercury, and oil are generally treated as incompressible, which means scientists assume the density is constant. Gases, such as air can, under certain limited circumstances, be treated as incompressible but generally require analyses that include compressibility—that is, changing density. It is for this reason that scientists developed and frequently apply the perfect gas equation of state.

Scientists generally start their work in fluid dynamics by invoking one, if not all, of the following assumptions: steady flow, incompressible flow, nonviscous flow, and irrotational flow. The condition of steady flow assumes that at any fixed point, the velocity of flowing fluid does not change in time. A lazy, gently wandering stream is a good example of this type of steady flow. Scientists describe this gentle, steady flow as laminar flow. In contrast, if the velocity (magnitude or direction) of a fluid is constantly changing at fixed points all along the flow path, scientists say the flow is not steady and describe the condition as turbulent flow. A

### PERFECT GAS EQUATION OF STATE

The perfect (ideal) gas equation of state is an important principle in fluid science, especially in the treatment of gases. This principle states that the pressure (P), absolute temperature (T), and volume (V) of a gas are related as follows: $P V = N R_U T$, where N is the number of moles of gas and $R_U$ is the universal gas constant. At low pressures and moderate temperatures, many real gases approximate perfect gas behavior very well. The important equation evolved after a century of careful, independent experimental work by the Irish-British scientist Robert Boyle (1627–91), the French physicist Jacques Charles (1746–1823), and the French chemist Joseph-Louis Gay-Lussac (1778–1850).

rapidly flowing stream often exhibits turbulent flow behavior, in which the flow of the water is dominated by eddies, apparent randomness, and recirculation.

The Irish-born British engineer Osborne Reynolds (1842–1912) investigated the flow of fluids, especially in pipes. He developed an important dimensionless ratio (now called the Reynolds number) to characterize a fluid's dynamic state. The Reynolds number ($N_{Re}$) predicts changes in fluid flow regimes (that is, laminar flow, transition flow, or turbulent flow). It is defined as the ratio of inertia force (momentum force) to viscous force in fluid flow. Engineers often use the following mathematical expression for the Reynolds number: $N_{Re} = (\rho\, l\, v)/\mu$, where $\rho$ is the fluid density, $l$ is the characteristic length of the fluid system, v is the fluid velocity, and $\mu$ is the absolute viscosity. At a low Reynolds number, the viscous force dominates, and the fluid flow and the motion is laminar. At high a Reynolds number, the inertia force dominates, leading to a condition of turbulent flow.

When scientists assume incompressible flow, they assume that the density of the fluid remains uniform and constant. This approximation is usually very good in treating flowing liquids but does not usually apply to flowing gases. The assumption of nonviscous (or inviscid) flow implies that the flowing "perfect" fluid has a zero coefficient of viscosity. Engineers often treat the flow of water as nonviscous (at least as a first approximation), but they must treat the flow of honey out of a jar or crude oil through a pipeline as viscous. A pipeline is a network of connected pipes assembled to transport a fluid, such as water, crude oil, or natural gas. The final approximation that scientists make in dealing with flowing fluids is the assumption of irrotational flow. This assumption neglects any internal swirling or twisting a fluid may experience as it flows down a channel or through a conduit.

## BERNOULLI PRINCIPLE

The Swiss mathematician Daniel Bernoulli (1700–82) became an important pioneer in fluid science when he published the classic work *Hydrodynamica* in 1738. While exploring the relationship between pressure, density, and velocity in flowing fluids, he discovered that a moving fluid exchanges its kinetic energy for pressure. Scientists now call this important observation the Bernoulli principle.

To appreciate the true significance of Bernoulli's amazing insight, consider first the steady state flow of a perfect (nonviscous and incompressible) fluid through a section of pipe of varying geometries, as shown in

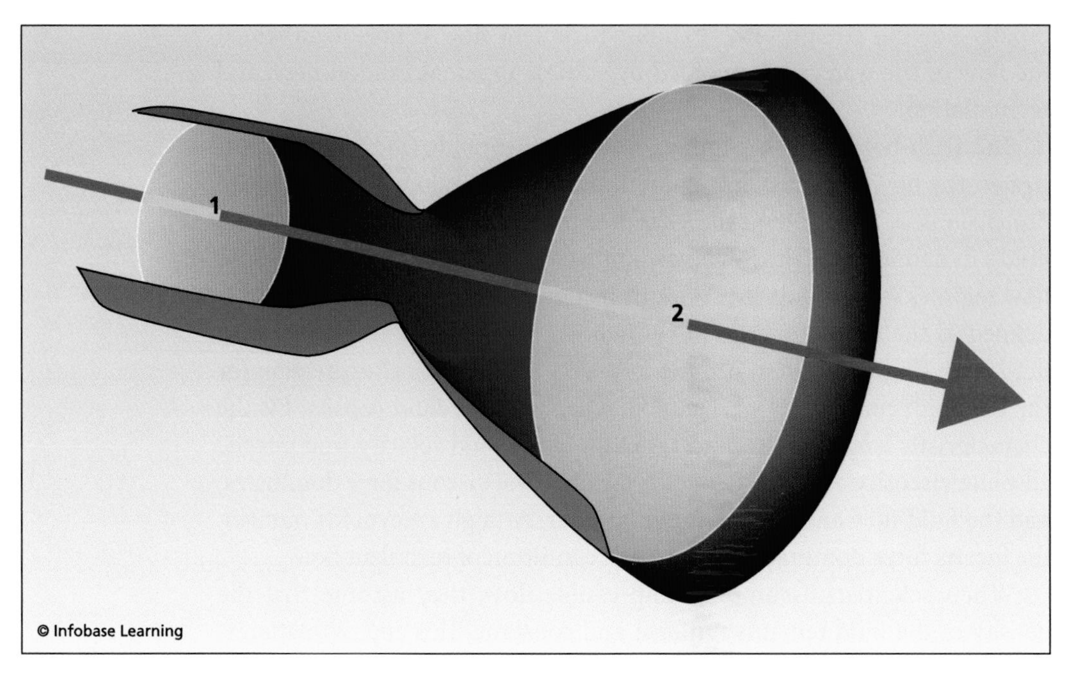


Basic diagram depicting fluid flow through a variable area conduit. In this diagram, section 1 designates the inlet or entrance; section 2 designates the outlet or exit. *(Modeled after NASA)*

the accompanying illustration. Since the flow is assumed steady, the mass flow rate ($\dot{m}$) at section 1 equals the mass flow rate at section 2. Scientists define mass flow rate ($\dot{m}$) using the following equation: ($\dot{m}$) = ρ A v, where ρ is the fluid's density, A is the cross-sectional area of the pipe or conduit, and v is the fluid velocity. For an incompressible (perfect) fluid, the continuity equation tells scientists that the product of the fluid's speed and pipe's cross-sectional area is a constant. Scientists express this very important relationship in fluid dynamics as: $A_1v_1 = A_2v_2$.

Applying Bernoulli's brilliant observations to the flow of a liquid such as water through the section of variable geometry pipe described in the accompanying figure, scientists can write the following: $[½ \rho v^2 + P]_1 = [½ \rho v^2 + P]_2$ = constant. As a consequence of Bernoulli's principle, if the velocity of the fluid increases, its pressure decreases, and vice versa.

This section has probably introduced enough equations and approximating assumptions to make most readers somewhat uncomfortable. In truth, predicting the flow of real fluids that behave in nonideal conditions is very complicated. Today, most scientists and engineers resort to computational fluid dynamics (CFD) to get the job done right.

## COMPUTATIONAL FLUID DYNAMICS

Computational fluid dynamics (CFD) is an exciting new branch of science that integrates the use of numerical techniques, high-speed computers, property data, and complex theoretical and empirical equations to generate high-quality approximate solutions for some of the most challenging fluid science problems. Central to many CFD activities are the Navier–Stokes equations. Claude Louis Marie Henri Navier (1785–1836) was a French civil engineer who specialized in mechanics. The Irish mathematician Sir George Gabriel Stokes (1819–1903) investigated the flow of viscous fluids. The Navier–Stokes equations are based on Newtonian physics and describe the motion of fluids for a variety of interesting cases.

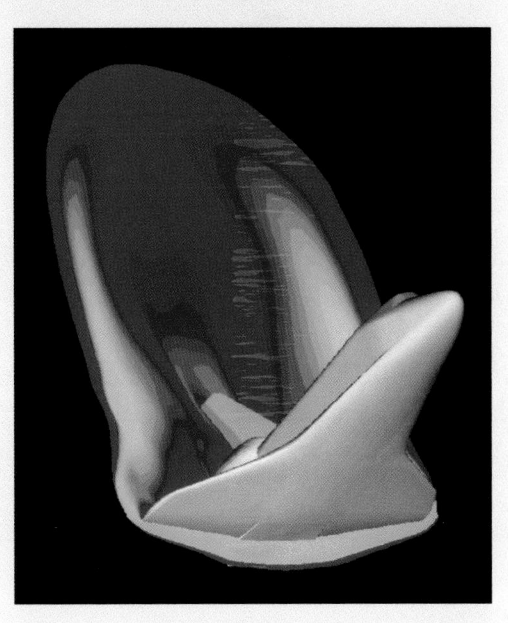

This is a computational fluid dynamics (CFD) computer-generated space shuttle orbiter vehicle model. *(NASA)*

The accompanying illustration is a computer-generated CFD model of NASA's space shuttle orbiter vehicle. This particular model was developed by NASA engineers during the early 1990s. As computing power increases and computer models become even more sophisticated, aeronautical and aerospace engineers are using CFD to supplant expensive wind tunnel experiments during the design of new flight vehicles. A wind tunnel is a ground facility that supports aerodynamic testing of aircraft, missiles, and propulsion systems under simulated flight conditions.

## PRANDTL AND BOUNDARY LAYER THEORY

Scientists define the boundary layer as the layer of fluid in the immediate vicinity of a bounding surface. The German physicist Ludwig Prandtl (1875–1953) revolutionized fluid science when he presented his boundary layer concept in 1904. His historic publication was entitled "On the motion

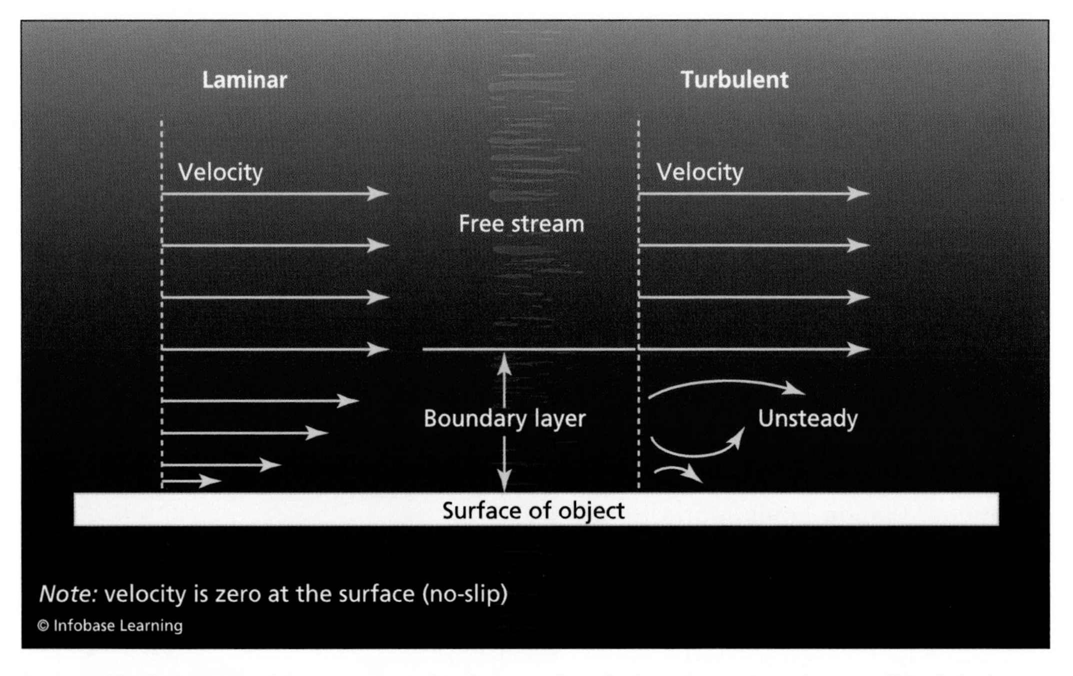


A simplified diagram that compares laminar and turbulent boundary layers *(Modeled after NASA)*

of fluids of very small viscosity." He discovered that while the bulk of a flowing fluid could be adequately treated using classical potential flow techniques, there was a thin region near an object where viscous effects arose. His work led to a much better understanding of skin friction, a contribution of great importance to the emerging aeronautics industry.

In fluid science, the boundary layer corresponds to the layer of fluid affected by the viscosity of the fluid. Within this thin layer of fluid near the surface, the fluid velocity changes from its free-stream velocity value to zero (at the surface or wall). Engineers often measure the thickness of the boundary layer from the surface to the point at which the fluid flow has 99 percent of its free-stream velocity. Boundary layers can be either laminar (layered flow) or turbulent (disordered), depending upon the Reynolds number. A thorough understanding of boundary layer phenomena is very important in fluid science. Unfortunately, the physical and mathematical details of boundary layer theory require an upper undergraduate or graduate school–level treatment, which far exceeds the scope or purpose of this book.

## SURFACE TENSION AND CAPILLARITY

Almost since the dawn of civilization, people have enjoyed the frolicking behavior of water in fountains. Scientists attribute the ability of water and other liquids to form streams and drops to a special physical property called surface tension. They define surface tension as the tendency of the cohesive intermolecular forces within a liquid to keep the liquid's surface as small as possible. This results in the formation of spherical drops and liquid streams. Bubbles are another manifestation of surface tension. The pressures inside and outside a bubble are dependent upon the surface tension of the bubble-forming liquid and the bubble's radius.

Capillarity is a closely related phenomenon. Capillary action causes a liquid to rise vertically, against gravity, up a thin tube or through small openings. The phenomenon is caused by a combination of adhesion and surface tension. Adhesion involves the mutual attraction that can take place between the molecules of a liquid and a solid container wall. Adhesion causes certain liquids, including water, to creep up the solid wall of a container a certain distance; surface tension helps the surface of the liquid form an upward curving (concave) feature called a meniscus. When the adhesive force between the solid wall and the liquid is greater than the cohesive force within the liquid, scientists say the fluid *wets* the surface of the solid. Capillarity helps plant draw water up from the soil through their roots. Capillary action is also responsible for a portion of the movement of groundwater.

If the cohesive forces within a liquid are greater than the adhesive forces between the liquid and the solid container wall, the liquid does not wet the container. For example, mercury does not wet a glass container. Rather, its surface forms a downward curving (convex) meniscus.

## NATURE'S MOST IMPORTANT PUMP

Engineers describe a pump as a machine that applies mechanical energy to a liquid flowing through it. The application of mechanical energy by the pump increases the pressure of the liquid, increases the rate at which the liquid is flowing, or both.

The human heart is nature's most magnificent pump. A person's heart is a muscular organ that acts like a pump. During life, it operates continuously, sending blood throughout the body. The British physician William Harvey (1578–1657) was the first scientist to describe the circulatory system and the central role played by the heart.

The human circulatory system consists of a network of blood vessels that includes arteries, veins, and capillaries. Capillaries are very tiny (microscopic) blood vessels that connect the smallest arteries (arterioles) and smallest veins (venules) throughout the body. Of all the blood vessels, medical scientists recognize that only capillaries have walls thin enough to permit the passage of the nutritive materials and oxygen in the blood into body tissues. Capillaries also absorb carbon dioxide and waste matter

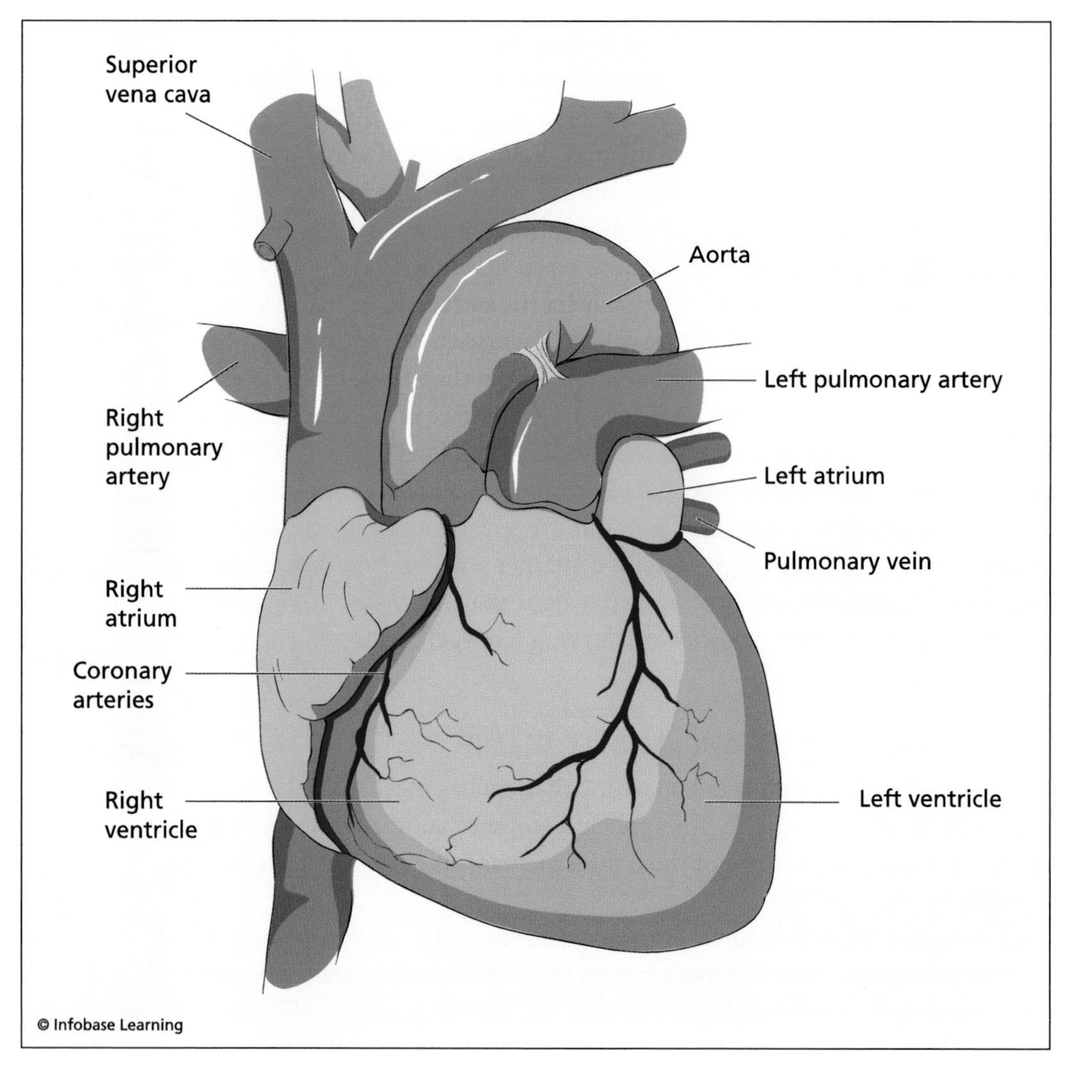


This illustration shows the front surface of a human heart, including the coronary arteries and major blood vessels. *(Modeled after NIH)*

from the tissues and place these in the bloodstream. The body's network of blood vessels carries blood to and from all areas of a person's body.

A typical adult human body contains between 1 and 1.6 gallons (3.8 to 6.1 L) of blood. Plasma constitutes up to 55 percent of human blood, with blood cells making up the remaining 45 percent. Finally, blood plasma contains about 91 percent water, 8 percent organic compounds, and 1 percent inorganic materials.

An electrical system regulates a person's heart and employs electrical signals to contract the heart's walls. When the heart's walls contract, blood is pumped into the body's circulatory system. There is a system of inlet and outlet valves within the heart's chambers that work together to ensure that blood flows in the proper direction.

The accompanying illustration shows the outside of a normal, healthy human heart. The heart is a muscle with four chambers: the right and left atria (in purple) and the right and left ventricles (in red). Some of the main blood vessels in the blood circulatory system connect directly to the heart. The right ventricle pumps oxygen-poor blood from the heart to the lungs. The left atrium receives oxygen-rich blood from the lungs. The pumping action of the heart's left ventricle sends oxygen-rich blood through the aorta (a main artery) to the rest of a person's body. Oxygen-poor blood from the vena cavae flows into the heart's right atrium and then on to the right ventricle. Healthy heart valves open and close in precise coordination with the heart's atria and ventricles. Each valve possesses a set of flaps, called leaflets or cusps, that seals or opens the valve. This intricate design allows pumped blood to pass through the heart's chambers and into a person's arteries without backing up or flowing backward.

CHAPTER 4

# Water—Nature's Most Amazing Molecule

**This chapter explains why scientists regard water as nature's miracle** liquid. The continuous movement of water between Earth's surface and atmosphere is discussed. Scientists call this natural process the hydrologic cycle, or the water cycle. The water cycle is the continuous process through which water is purified by evaporation and transported back to Earth's surface. Hydrology is the branch of science that investigates the properties of water; the occurrence, distribution, and movement of water on Earth; and the relationship of moving water to the environment during each phase of the hydrologic cycle.

## PHYSICAL AND CHEMICAL PROPERTIES OF WATER

Water is necessary for all living things. Pure water is virtually colorless, odorless, and tasteless, yet its physical and chemical properties make it nature's most amazing molecule. Many people recognize the water molecule's well-known chemical formula, namely $H_2O$. What most people do not appreciate, however, is how this apparently simple molecule can possess so many intriguing qualities and hidden properties.

The accompanying diagram presents a ball-and-stick model of the water molecule. As depicted, each water molecule contains one atom of oxygen (colored red) bound to two atoms of hydrogen (colored white). The hydrogen atoms are covalently bonded (attached) to one side of the oxygen atom. (A covalent bond is the chemical bond created within a molecule when two or more atoms share an electron.) This lopsided arrangement results in a bent molecule that has a positive charge on the side, where the hydrogen atoms are, and a negative charge on the other side, where the oxygen atom is. Specifically, because the bonding electrons are shared unequally by the oxygen atom and hydrogen atoms, two partial negative charges (symbolized as $\delta^{2-}$ in the illustration) form at the oxygen end of the water molecule, while a partial positive charge (symbolized as $\delta^{+}$) occurs at the hydrogen ends of the molecule.

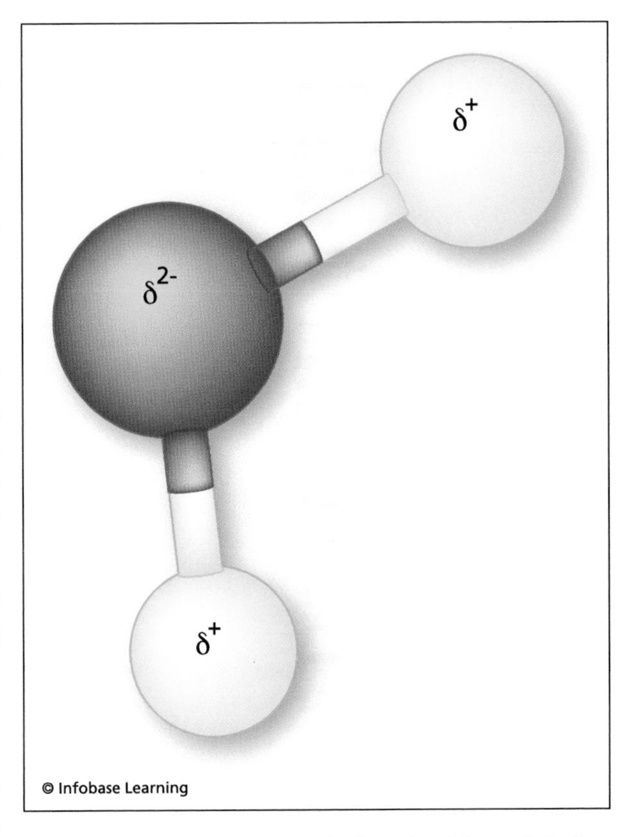


This illustration presents a ball-and-stick model of the highly polar water molecule. The oxygen atom is red, and the hydrogen atoms are white. The bent molecule has an experimentally measured angle of 104.5°. The symbol δ represents partial charge in a polar molecule. *(Modeled after DOE/PNNL)*

This molecular dipole arrangement causes individual water molecules to interact with one another. Since oppositely signed electrical charges attract, the side of a water molecule with the hydrogen atoms (the $\delta^{+}$ partial charges) attracts the oxygen side (the $\delta^{2-}$ partial charge) of an adjacent water molecule. This attractive behavior allows water molecules to stick together and form macroscopically observable drops. Scientists refer to this particular type of intermolecular force as hydrogen bonding. Although relatively weak compared to the H-O covalent bonding that occurs between hydrogen and oxygen atoms within an individual water molecule, hydrogen bonding creates the intermolecular force between adjacent water molecules that produces many of water's intriguing properties.

One especially unusual and important property of water is that water's solid form, ice, is less dense than its liquid form at 32°F (0°C). As it freezes, water expands. The fact that ice floats on water has profound implications for life on Earth.

A close-up look at a drop of nature's miracle liquid, water *(DOE/FNAL)*

Another significant property of water is its unusually high value of specific heat capacity. Scientists and engineers define the specific heat capacity of a substance (c) as a measure of that substance's ability to store energy received as heat per unit mass. Basically, it is the quantity of heat needed to raise the temperature of a unit mass of substance by one degree. Before discussing the implication of water's high specific heat capacity, it is quite helpful to understand the origin and significance of the British thermal unit (Btu).

Early in the 19th century, during the formative years of classical thermodynamics, British engineers needed a way to measure and compare energy transfer as heat. They defined the British thermal unit (Btu) as the amount of heat needed to raise the temperature of one pound-mass (lbm) of water one degree Fahrenheit (°F). Some of the conditions surrounding this definition required that the water be initially at 59.5°F (15.3°C) and one atmosphere pressure. On the European continent, other scientists developed an early metric equivalent called the calorie (cal). They defined one calorie as the amount of heat required to raise one gram (g) of water at one atmosphere pressure from 14.5°C (58.1°F) to 15.5°C (59.9°F). Scientists and engineers no longer use the calorie to express thermal energy units, except in discussions of the energy content in kilocalories of food. Today, by international agreement, one British thermal unit is equal to 1,055 joules (J), where the joule is the basic unit of energy or work in the SI unit system.

Returning to the discussion of specific heat, water has a value of 1.00 Btu/(lbm–°F) (4.18 kJ/kg–°C) at a constant pressure of one atmosphere and a temperature of 77°F (25°C). The element mercury (Hg), a liquid under these conditions of pressure and temperature, has a specific heat of only 0.0334 Btu/(lbm–°F) (0.14 kJ/kg–°C). This means it takes about 30 times as much heat to raise the temperature of a unit mass of water one degree as it does to raise an identical mass of mercury one degree when both substances start at the same temperature and pressure.

Water's high value of *heat of vaporization* is another unusual and important property. Scientists and engineers prefer to use the term *enthalpy of vaporization,* where the specific enthalpy (h) is an important thermodynamic property that does not lend itself to a simple physical interpretation. The essential point here is that a large amount of heat is needed to evaporate a small quantity of water. Conversely, in order for a small quantity of water to condense from a vapor (gaseous state) to a liquid, it must release a large amount of heat. Thermodynamic property data indicate that at a temperature of 212°F (100°C) and a pressure of one atmosphere, the heat of vaporization (enthalpy of evaporation) of saturated water is 970.35 Btu/lbm (2,257.03 kJ/kg).

About 75 percent of Earth's surface is covered with water. How all this surface water interacts with incoming sunlight depends to a large extent on the unusual properties of the water molecule. For example, water's high values of specific heat capacity and heat of vaporization exert a great influence on the transport and distribution of energy throughout Earth's atmosphere as part of the hydrologic cycle (discussed in the next section).

As both a polar substance and a hydrogen-bonding molecule, water exhibits unusual solvent properties. Many scientists refer to water as the "universal solvent" because it dissolves more substances than any other liquid. This means that everywhere water journeys, whether it is within a person's body or through the ground, it collects and transports chemicals, minerals, and nutrients. This process is essential for life on Earth.

Environmental scientists point out that the amazing solvent properties of water also make water a highly mobile carrier of dissolved pollutants and waterborne diseases. Throughout recorded history, people perceived "dilution as the solution." Human wastes along with toxic materials were often mixed with large quantities of water in the hope of making the problem "disappear" within large natural bodies of water, especially the oceans. However, once dangerous substances dissolve in water, they can more easily disperse throughout the environment, accumulate in food chains, and cause widespread harm. Modern cleanup activities aimed at neutralizing or removing hazardous water-dissolved and -dispersed substances have often proved technically challenging as well as very expensive. Furthermore, many strains of bacteria and microorganisms harmful to human beings thrive in water. Water pollution is a global problem.

Another unique feature of water is that it is the only natural substance found in all three physical states, solid (ice), liquid, and gas (steam), at the temperatures and pressures normally found on Earth's surface. Throughout the hydrologic cycle (discussed in the next section), Earth's water supply constantly experiences movement, interactions, and changes of state.

## SOLUTIONS

When scientists dissolve a substance in a pure liquid, they refer to the dissolved substance as the solute and the host pure liquid as the *solvent.* They call the resulting intimate mixture the solution. Generally, there is more solvent than solute.

Scientists define an aqueous solution as one in which water is the solvent. Saline solution is an aqueous solution commonly used in most hospitals. Pharmacy technicians prepare this solution by dissolving a certain mass of sodium chloride (NaCl), a solid substance more commonly called salt, with a certain mass of water. For example, the normal (or isotonic) saline solution found in hospitals contains approximately 0.9 percent sodium chloride by mass, with the remainder of the intravenous solution's mass consisting of pure water. An isotonic saline solution is one that has the same salt concentration as the normal cells of the body and the blood. Athletes often drink isotonic beverages to replace the fluid and minerals their bodies expend during strenuous physical activities.

In many practical applications, people express solution concentrations in percent by volume. Here, the percent by volume of the solution equals the volume of the solute divided by the total volume of the solution. The topical antiseptic rubbing alcohol is a water-isopropyl alcohol [$C_3H_8O$ or $(CH_3)_2CHOH$] solution that typically contains about 70 percent isopropyl alcohol by volume.

Chemists generally prefer to use molarity *(M)* as a concentration unit in dealing with solutions. Within the SI unit system, they define molarity (or

Pure water is neither acidic nor basic. Rather, it has a neutral pH value of 7. The Danish biochemist Søren Peder Lauritz Sørensen (1868–1939) proposed the pH scale in 1909. As shown in the illustration on page 54, pH values range from 0 to 14, with 7 representing a neutral value. The pH scale presents values on a logarithmic basis. The *H* in pH represents hydrogen, and the *p* means power. Pure water has a molar concentration *(M)* of hydrogen ions ($H^+$) and hydroxyl ions ($OH^-$) of $1.0 \times 10^{-7}$. Following in Sørensen's footsteps, chemists define pH as the negative logarithm (to the base 10) of the molar concentration of hydrogen ions—that is, $pH = -\log_{10}[H^+]$, where the brackets around $H^+$ symbolize molarity. This means that each unit on the pH scale expresses a 10-fold change in how acidic or basic a substance is. For example, water with a pH of 4 is 10 times more acidic than water having a pH of 5. The chart also depicts the pH values molar concentration) as the amount of solute (expressed in moles) per liter of solution. When laboratory technicians want to prepare a solution that is 0.200 *M* $CuSO_4$, they first add a modest amount of water to dissolve 0.200 moles of copper sulfate ($CuSO_4$) crystals placed in the bottom of a volumetric beaker. Next, they add more water, until the solution reaches the one liter (1 L) mark on the beaker. Finally, they stir until they create a uniformly mixed solution.

In dealing with solutions, two qualitative terms are often encountered: dilute solution and concentrated solution. Chemists identify a dilute solution as one that has little solute compared to the solvent; while a *concentrated solution* has a relatively large amount of solute compared to the solvent.

A U.S. Air Force medical technician squeezes drops of a patient's blood into a glass beaker containing copper sulfate ($CuSO_4$) solution. If the blood does not descend to the bottom of the beaker in 15 seconds, the patient has an iron deficiency. *(USAF)*

of several common solutions as well as some of the consequences of acid rain.

Scientists define an acid as a substance that produces or donates hydrogen ions ($H^+$) when dissolved in water, while a *base* is a substance that, when added to water, produces an excess of hydroxide ions ($OH^-$). In other words, a base is a proton acceptor. Therefore, pH is a measure of the relative amount of free hydrogen and hydroxyl ions in water.

Highly concentrated (1 *M*) hydrochloric (HCl) acid contains $6.02 \times 10^{23}$ hydrogen ions ($H^+$) per liter of water and corresponds to a pH value of 0. Chemists place highly concentrated (1 *M*) sodium hydroxide (NaOH) at the other end of the pH scale and assign the extremely basic (alkaline) solution a pH value of 14. Here are the approximate pH values for several other common solutions. Orange juice is about 3.5, human urine varies from 5 to 7, milk is about 6.5, human blood is approximately 7.5, seawater ranges from 8.5 to 10, and household ammonia is approximately 11.5.

Environmental scientists regard a changing pH value for the water in a stream, river, or lake as an indicator of pollution. The pH of water deter-

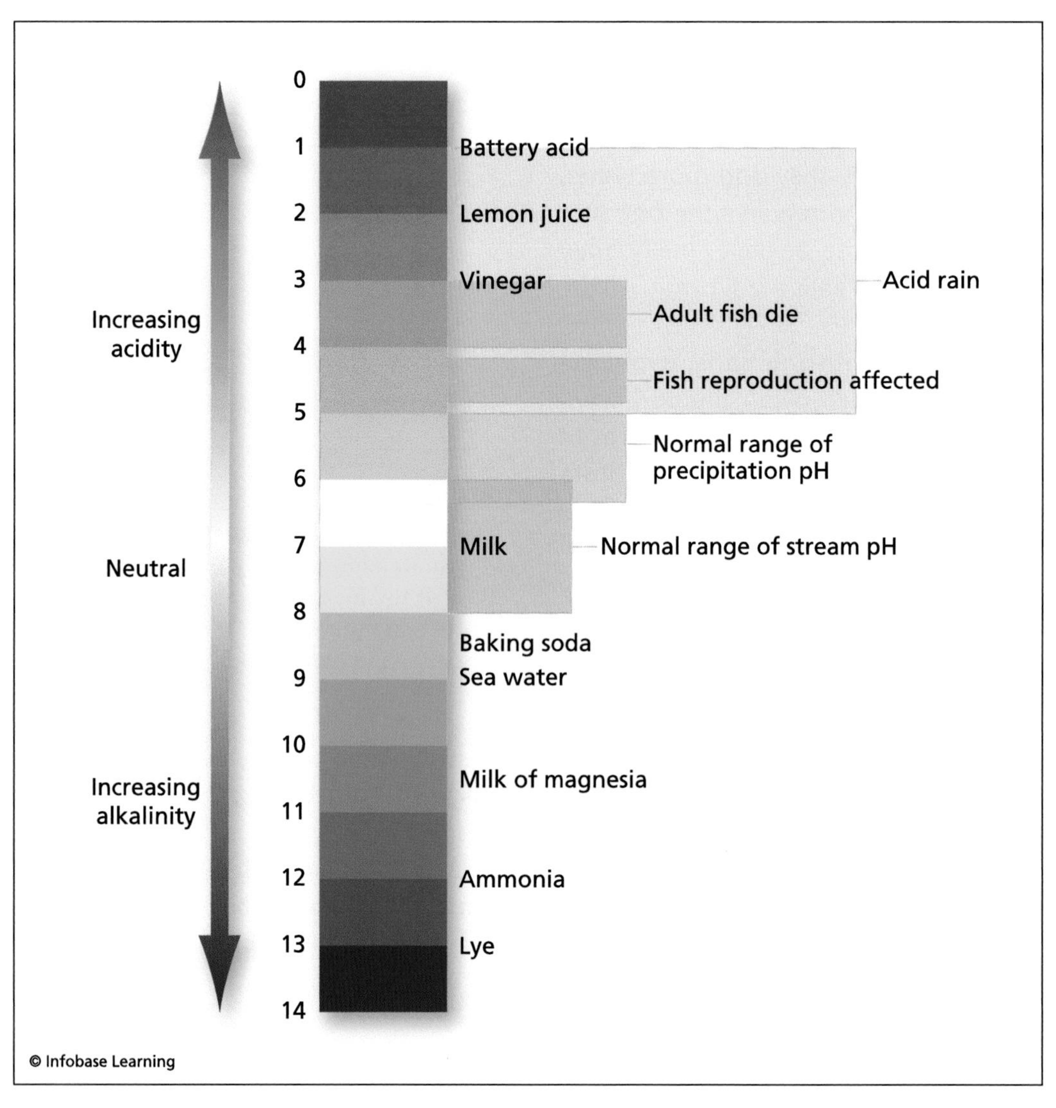


Chart of pH ranges. Water with pH values less than 7 is acidic, while water with pH values greater than 7 is basic (or alkaline). Normal precipitation (rainfall) has a pH value of about 5.6, which is slightly acidic due to carbon dioxide gas from the atmosphere. Acid rain represents very acidic water, which can cause adverse environmental consequences. *(Modeled after USGS and Environment Canada)*

mines the solubility and biological availability of chemical constituents, including nutrients such as phosphorus, nitrogen, and carbon and heavy metals such as lead, copper, and cadmium. Scientists know that the degree to which heavy metals are soluble establishes their toxicity. These metals tend to be more toxic at lower pH because they are more soluble under such conditions. Environmental biologists recognize that pH level determines how much and what form of phosphorous is most prevalent in a particular body of water. They use pH value as an indicator of what forms of aquatic life, if any, can survive in that particular body of water. For example, if a freshwater lake has a pH value of 4 or less, adult fish will probably die.

## EARTH'S HYDROLOGIC CYCLE

Earth's great hydrologic, or water, cycle consists of the continuous journey made by water molecules from the planet's surface, into the atmosphere, and back again. The hydrologic cycle has been operating for billions of years, and life on Earth depends on it. Radiant energy from the Sun powers this planetary process, which involves the continuous exchange of moisture between the oceans (discussed in chapter 5), the atmosphere, and the land. As depicted in the illustration, there are many interesting components to Earth's water cycle. This section focuses upon those that scientists consider the most important: evaporation and transpiration, condensation, precipitation, and runoff and groundwater.

During the hydrologic cycle, radiant energy from the Sun causes large quantities of water to evaporate from the surface of the oceans and other bodies of water, such as streams, rivers, and lakes, and enter the atmosphere. Water can also evaporate from moist soil. Scientists define evaporation as the physical process by which a liquid transforms into a gas (vapor) at a temperature below the boiling point of the liquid. Scientific studies indicate that the oceans, seas, and other bodies of surface water provide 90 percent of the moisture in Earth's atmosphere.

A little less than 10 percent of the moisture in Earth's atmosphere is released by plants, trees, and other vegetation through the process of transpiration. Plants and trees collect water through their root systems as part of the delivery of nutrients to their leaves. They then release this water back into the atmosphere by transpiration, which is basically how plants sweat and cool themselves. In fluid mechanics, engineers describe transpiration *cooling* as a form of mass transfer that involves controlled injection of a fluid mass through a porous surface. This process is basically limited by

## ACID RAIN

The expression *acid rain* is a broad, general term that refers to a mixture of wet and dry deposition of acidic materials from the atmosphere. By causing freshwater streams and lakes to become more acidic (that is, to have lower pH values), acid raid adversely affects ecological systems.

Raindrops descending through the air can absorb carbon dioxide ($CO_2$) and become a dilute solution of carbonic acid ($H_2CO_3$), a weak acid. Rainwater that has become saturated with carbon dioxide has a natural pH value of 5.6, but in many parts of the world, rainfall that occurs downwind of industrial sites can become much more acidic, acquiring a pH of 3 or less, due to interaction with human-caused pollutants. These anthropogenic pollutants include sulfur dioxide ($SO_2$) from the burning of high–sulfur content coal in electrical generating plants and nitrogen dioxide ($NO_2$) and nitric oxide (NO) from automobile emissions. The smoke and fumes that accompany the burning of fossil fuels ascend high into the atmosphere and combine with moisture in the air to form mild solutions of sulfuric acid ($H_2SO_4$) and nitric acid ($HNO_3$). Precipitation (rain, snow, and fog) brings these pollutants back to Earth as acid rain. Scientists use the term *wet deposition* to describe acidic rain, snow, fog, and mist.

As this acidic water flows over and through the ground, it can adversely affect a wide variety of plants and animals. The magnitude and extent of the environmental consequences depend on several factors, including the acidity (pH value) of the precipitation, the chemistry of the soil, and the ecological systems being stressed. Scientists at the Environmental Protection Agency (EPA) estimate that about 67 percent of all sulfur dioxide ($SO_2$) and 25 percent of all nitrogen oxides ($NO_x$) emissions in the United States come from electric power generation that relies on the combustion of fossil fuels such as coal.

In dry deposition, some of the human-produced acidity returns to Earth from the atmosphere as particles and gases. As the wind blows caustic particles and acidic gases into contact with trees, buildings, homes, and automobiles, the objects they encounter often deteriorate. Sometimes, rainstorms combine acid

the maximum rate at which the coolant material can be pumped through a porous surface.

Two other natural processes send water from Earth's surface back into the atmosphere. The first process involves sublimation from snowbanks rain with dry-deposited acid, leading to aggravated cases of acid deposition and severe environmental consequences. Natural phenomena such as volcanic eruptions and lightning also add acid rain precursor pollutants into the atmosphere. Volcanoes release sulfur dioxides and sulfuric acid, while lightning generates nitrogen oxides and nitric acid.

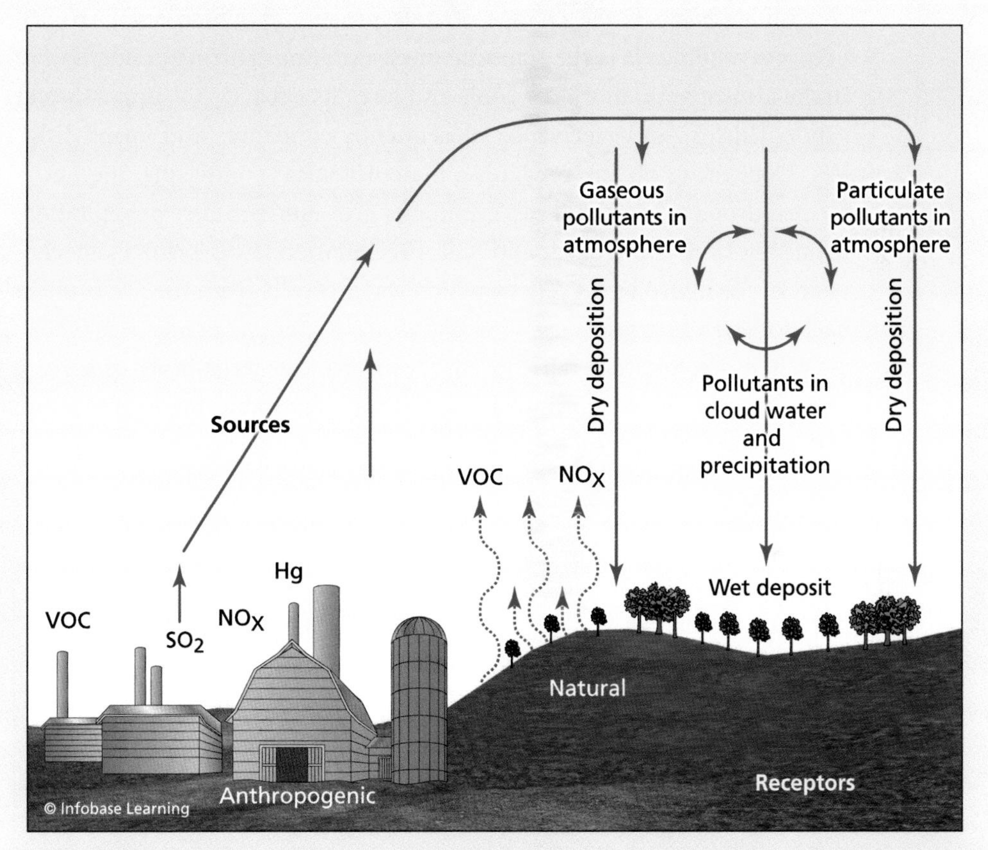


**Formation of acid rain by human-generated sulfur dioxide ($SO_2$) and nitric oxide ($NO_x$) emissions into the atmosphere. Other atmospheric pollutants include mercury (Hg) and volatile organic compounds (VOCs).** ***(Modeled after EPA)***

and ice fields. Scientists describe sublimation as the direct transition of a substance from the solid state (here, ice or snow) to the gaseous state (here, water vapor) without passing through the liquid state. Volcanic eruptions (see chapter 9) also inject water vapor and other gases into the atmosphere.

As air containing water vapor rises into the atmosphere, it expands and cools. When the ascending moist air cools to a certain temperature, called the dew point, it becomes a saturated vapor and begins to condense. Scientists define the dew point as the temperature at which water vapor condenses to liquid water when it is cooled at constant pressure. Clouds, fog, dew, mist, and frost are examples of condensation in the atmosphere.

Condensation is the change of state process by which a vapor (gas) becomes a liquid. It is the opposite of evaporation. During condensation in the atmosphere, moist air cools and loses its capacity to support water vapor. This excess water vapor subsequently condenses and forms droplets. Depending on the prevailing meteorological conditions, the newly formed droplets in a cloud can eventually grow and produce precipitation, including rain, snow, sleet, and hail. Precipitation is the primary mechanism in the hydrologic cycle by which water travels from the atmosphere back to Earth's surface.

When precipitation falls on an ocean or other large body of surface water, it gets stored and awaits conditions favorable for evaporation.

Basic components of Earth's great hydrologic (water) cycle *(Modeled after NOAA)*

When precipitation falls over land, water follows one of several significant routes. Some precipitation falls as snow and accumulates in ice caps or glaciers, both of which can store frozen water for thousands of years. As climate conditions change, the ice caps or glaciers can grow and store more frozen water or else shrink and retreat, melting in the process and returning the snowmelt runoff to streams and ultimately the oceans. An ice cap, glacier, or snowpack may also return a small quantity of moisture directly to the air through sublimation. In warmer (nonarctic) climates, the annual thawing and melting of snowpacks releases snowmelt that flows overland toward oceans or large lakes. In many parts of the world, the snowpack represents an important water resource, which annually feeds local streams and rivers as it melts. However, sudden melting of a snowpack due to unseasonably warm weather late in winter can cause severe flooding.

Generally, when precipitation falls on land, it will either be absorbed into the ground, eventually becoming groundwater, or, if the ground cannot absorb any more water, it becomes runoff and flows into streams under the influence of gravity. Some of the groundwater is taken up by plants and trees and returns to the atmosphere through transpiration, while some of the runoff evaporates as it flows into streams or rivers. Part of the precipitation that infiltrates the ground eventually make its way into streams. The water in streams then converges into rivers and flows back to the oceans, which contain about 97 percent of the planet's total supply of water. According to scientists at the United States Geological Survey (USGS), the remaining 3 percent of the planet's water supply is freshwater. Of this amount of freshwater, about 69 percent is locked up in icecaps and glaciers, 30 percent is in the ground, and less than 1 percent resides in surface water sources. Finally, lakes contain about 87 percent of the world's freshwater sources, swamps about 11 percent, and rivers about 2 percent.

When hydrologists examine the amount of water that flows in rivers, they use the concept of the watershed. They define a watershed as the area of land where all the water that precipitates upon it and drains off it travels to the same place. A watershed consists of surface water, such as streams, lakes, wetlands, and reservoirs, as well as all the underlying groundwater. A very large watershed such as Chesapeake Bay consists of many smaller watersheds and typically has an ocean or large lake at its outflow point. Natural processes as well as water use by people upstream of the outflow point can significantly affect the quality and quantity of water that departs a particular watershed.

In most cases, when precipitation falls on the surface of the land, a portion of the water runs off into rivers and lakes. Another portion of the rainfall seeps into the soil. Hydrologists call the process by which rainwater enters the soil infiltration and define infiltration rate as the rate at which soil can absorb rainfall or irrigation. They usually measure this important quantity in inches of water per hour (cm/hr). Runoff occurs whenever the precipitation or irrigation rate exceeds the infiltration rate. Many factors influence the infiltration process, including the soil's porosity and permeability, the rainfall intensity, and the vegetation cover. Land surfaces covered with forests have high infiltration rates. It is through infiltration that groundwater begins its life on the surface of the land.

Since large amounts of water are stored in this manner, groundwater plays a major role in the hydrologic cycle. The accompanying diagram shows that water beneath the land surface occurs in two major zones, the

## LAKES AND RIVERS

Lakes and rivers serve as the daily sources of freshwater for many people. A river consists of surface water journeying over land from a higher altitude to a lower altitude under the influence of gravity. Within the hydrologic cycle, runoff water flows into small streams, which merge to form larger streams and ultimately rivers. Most large rivers reach the oceans. On occasion, however, flowing surface water encounters a place surrounded by high terrain on all sides. When this happens, the flowing water naturally accumulates in the basin (or bowl) and forms a lake. For stable, aquatic life–supporting lakes, the water inflow each year balances the outflow, including seepage, surface drainage, and evaporation. The majority of lakes on Earth contain freshwater, but isolated lakes that lack proper drainage and outflow can develop a high salt content. One notable example is the Great Salt Lake in Utah. When engineers construct a dam to restrict a river's natural flow, they call the lake that forms behind the dam a *reservoir.*

Using surface area as the figure of significance, Lake Superior is the largest freshwater lake in the world. Situated in North America between the United States and Canada, this lake has a surface area of approximately 31,820 square miles (82,400 km$^2$), an average depth of 482 feet (147 m), and contains some unsaturated zone and the saturated zone. Scientists term the upper level of the soil the unsaturated zone because water there is present in varying amounts that change over time but not in sufficient quantities to saturate the soil. In the unsaturated zone, the voids within the rocks contain both air and water. The upper part of this zone is crisscrossed by active plant roots, animal and insect burrows, and voids left by decayed roots, all of which enhance the infiltration process. Hydrologists recognize that although a considerable amount of water may be present in the unsaturated zone, people cannot use wells to pump this water because it is too tightly held by capillary forces.

Below the unsaturated zone is a region called the saturated zone, where all the cracks, pores, and spaces between rock particles are saturated with water. Hydrologists use the term *groundwater* to identify water found in the saturated region. They also use the term *aquifer* to describe water-bearing underground formations capable of providing enough water to

2,900 cubic miles (12,100 km$^3$) of freshwater. The highest freshwater lake in the world that is commercially navigable is Lake Titicaca in South America. Located high in the Andes Mountains at an altitude of 12,507 feet (3,812 m), Lake Titicaca forms part of the international border between Peru and Bolivia. In contrast, the Dead Sea is the lowest lake in the world. Bordering Israel, Jordan, and the West Bank, the surface of the Dead Sea lies some 1,370 feet (418 m) below sea level. Although the lake is fed by the Jordan River and several small streams, it has no outlet, save for evaporation, so it is now one of the saltiest bodies of water in the world and contains no aquatic life.

Throughout history, rivers have played a critical role in the evolution and maintenance of civilizations. The Nile River in Africa is the longest river in the world, with an estimated length of 4,132 miles (6,648 km). It is followed closely by the Amazon River in South America, which has a length of 4,000 miles (6,436 km), and then the Yangtze River in Asia, which has a length of 3,915 miles (6,299 km). The two longest rivers in the United States are the Missouri River at 2,540 miles (4,087 km) and the Mississippi River at 2,340 miles (3,765 km). The Saint Lawrence River at 1,900 miles (3,057 km) long forms part of the international border between the United States and Canada, while the Rio Grande, also about 1,900 miles (3,057 km) long, creates a major portion of the border between the United States and Mexico.

Diagram that depicts water beneath the land surface *(Modeled after USGS)*

satisfy at least local human needs. Aquifers occur all around the globe and represent a major source of groundwater to support the daily needs of many people. The upper region of the saturated zone is called the water table. People can drill wells into an aquifer and pump water out of it, but if they pump too much water over too short a period of time, the water table drops and the wells run dry. The depth of the water table is highly variable. The journey of precipitation into and through the ground is a complicated process. There are many factors that determine the rate at which a particular aquifer gets recharged (replenished).

According to hydrologists at the USGS, groundwater represents the largest source of usable freshwater storage in the United States and contains more water than all surface lakes and reservoirs, including the Great Lakes. In many locations, groundwater represents the sole source for municipal water systems. However, groundwater can become polluted—an undesirable condition that is less visible than surface water pollution and more difficult to clean up. Some of the major sources of groundwater pollution are the improper dumping of industrial chemicals and wastes, leaky landfills that contain mixtures of urban and industrial trash, chemicals from mine tailings and wastewater ponds that seep into the ground, leaking underground oil and gasoline storage tanks, and broken or abandoned septic tanks.

CHAPTER 5

# The Oceans

**Viewed from outer space, Earth's most distinctive feature is the** majestic world ocean, which covers approximately 71 percent of the planet's surface and contains 97 percent of the planet's water. This chapter describes the five named oceans and explains their roles in the operation of Earth as a life-sustaining planet. The chapter also discusses how Earth's liquid-water ocean hosts nearly 50 percent of all species found on Earth and provides about 5 percent of the total protein in the human diet.

## PHYSICAL DESCRIPTION AND FEATURES

In Greek mythology, the Titans were a race of very powerful elder gods who ruled supreme over the primordial universe until they were overthrown by Zeus (Jupiter) and the other Olympian gods. The ancient Greeks assumed that a vast "world river" surrounded all habitable land. They even personified this world river and named it after the Titan Oceanus (Ωκεανος). The modern word *ocean* derives from this personification of Earth's large global body of salt water. The two major bodies of salt water known to the ancient Greeks and Romans were the Atlantic Ocean (primarily coastal portions adjacent to Europe) and the Mediterranean Sea. Greco-Roman mythology further elaborated that the Olympian god Poseidon (Neptune) ruled over the seas, especially the Mediterranean Sea. Other early peoples around the world developed mythologies and legends

Yellowfin tuna coming aboard during pole and line fishing operation in the tropical Pacific Ocean *(NOAA)*

to help explain and characterize those portions of the world ocean that influenced their lives.

Space age images of Earth have provided scientists and geographers a more precise way to measure the true extent of the vast, interconnected body of salt water collectively referred to as the world ocean. Although physically interconnected, people from ancient times to the present day have found it convenient to divide this world ocean into somewhat smaller, more easily comprehended bodies of salty water. In daily language, people often use the words *ocean* and *sea* interchangeably—for example, *ocean water* versus *seawater* or an *ocean cruise* versus a *sea cruise.*

Earth has a total surface area of 197.024 million square miles (510.072 million km$^2$). According to statistics presented in the *2009 World Factbook,* published by the Central Intelligence Agency, water covers 139.49 million square miles (361.132 million km$^2$), that is, 70.9 percent of the total surface area, and land covers 57.53 million square miles (148.94 million

## SALINITY

Scientists define salinity as a measure of the quantity (concentration) of dissolved solids (salts) in a specified quantity of water. Traditionally, the salinity value represented the amount of dissolved salts, expressed as either parts per thousand (ppt) or as mass of salt per reference mass of water. For example, open ocean water with a salinity of 35 ppt contains 35 pounds-mass of salt per 1,000 pounds-mass of salt water (35 kg of salt per 1,000 kg of salt water). Sodium chloride (NaCl), or common table salt, is the most abundant of the many types of salts present in ocean water.

By international agreement in 1978, scientists began to measure salinity in terms of a dimensionless quantity called the practical salinity unit (psu). Since the United Nations Educational, Scientific, and Cultural Organization (UNESCO) Practical Salinity Scale of 1978 (PSS78) defined salinity in terms of a referenced electrical conductivity ratio, the scientists made the measurement of salinity a dimensionless scientific quantity. Although scientists based the psu upon a more accurate and complex definition, salinity measurements expressed in psu and in ppt actually have nearly equivalent numerical values.

Most contemporary scientists express the salinity of water in terms of the psu—obtained from the conductivity-temperature-depth (CTD) instrument. This instrument provides in situ measurements of electrical conductivity (salinity), temperature, and pressure—physical parameters directly related to the density of water. By measuring conductivity, that is, how easily an electric current passes through the sample of water being tested, scientists obtain a direct indication of the water sample's salinity. The more salt that is dissolved in the water, the higher the electric conductivity. At one atmosphere pressure, freshwater freezes at 32°F (0°C), but the freezing point of ocean water varies with salinity. As a simple rule of thumb, for every 5 psu increase in salinity, the freezing point of water decreases by 0.5°F (0.28°C). Consequently, in the North Atlantic Ocean, where the salinity is about 35 psu, surface water begins to freeze at 28.8°F (–1.8°C). The freezing of seawater is a complex process in which salinity plays an important role.

Sea surface salinity (SSS) varies considerably due to hydrologic cycle processes such as evaporation and precipitation. In the open ocean, lower values of SSS generally occur where precipitation is greater than evaporation, and higher values of SSS occur where evaporation is greater than precipitation. For

*(continues)*

*(continued)*

certain ocean regions, SSS is significantly influenced by the inflow of freshwater from large rivers as well as the melting of polar region freshwater ice sheets and glaciers.

Distilled water contains no dissolved salts (psu = 0). Typical drinking water has a psu value of less than 0.1, and a salinity value of 1 psu delimits the maximum allowable level for potable drinking water. Brackish water has a salinity value ranging between 0.5 and 30 psu. The average salinity of water in the open ocean ranges from approximately 32 to 37 psu. Scientists refer to water with a salinity of more than 50 psu as brine.

km$^2$), that is, 29.1 percent of the total surface area. By historic tradition and recent international agreement, Earth's world ocean is divided into five bodies of water (in descending order of size): Pacific Ocean, Atlantic Ocean, Indian Ocean, Southern Ocean, and Arctic Ocean.

## IMPORTANCE OF THE OCEANS

What uniquely separates Earth from the other terrestrial planets (Mercury, Venus, and Mars) and the Moon is the ubiquitous presence of water as a liquid everywhere on humans' home planet. Water is the only known substance that can naturally exist in all three commonly encountered physical states (solid, liquid, and gaseous) within the relatively small range of pressures and atmospheric temperatures found at Earth's surface.

Scientists estimate that Earth contains about 331 million cubic miles ($1.38 \times 10^9$ km$^3$) of water. Of this huge quantity of water, approximately 96.5 percent is stored in the oceans. About 1.7 percent of the planet's water is stored in the polar icecaps, glaciers, and permanent snow. Another 1.7 percent is stored in groundwater, lakes, rivers, streams, and soil. While clouds are a common feature of the terrestrial atmosphere, less than 0.1 percent of Earth's total water inventory actually resides in the atmosphere as water vapor at any given moment. However, large quantities of water—some 91,000 cubic miles (380,000 km$^3$)—cycle through the atmosphere each year as part of Earth's hydrologic cycle. (The previous chapter discusses the hydrologic cycle.) Solar heating and wind action cause about 77,000 cubic miles (321,000 km$^3$) of water to evaporate as freshwater from

## SCIENTIFIC PROFILE OF THE OCEAN

Marine biologists and oceanographers commonly divide the ocean's life zones into two general regions, the coastal zone and the open sea. The coastal zone contains about 90 percent of all marine species. It is characterized by warm, nutrient-rich shallow water that extends from land (typically, the high-tide mark) to the edge of the gently sloping portion of the continental shelf. Deposits of nutrients on the ocean floor support numerous scavengers such as crabs, filter feeders such as oysters, and deposit feeders such as worms. The open sea begins with a dramatic increase in water depth at the end of the continental shelf.

The word *pelagic* means "of, pertaining to, or living in the open sea." As depicted in the accompanying illustration, scientists divide the open sea into several distinct zones: the epipelagic zone, mesopelagic zone, bathypelagic zone, abyssopelagic zone, and hadalpelagic zone. The epipelagic zone extends from the surface to a depth of about 660 feet (200 m). This zone is also called

*(continues)*

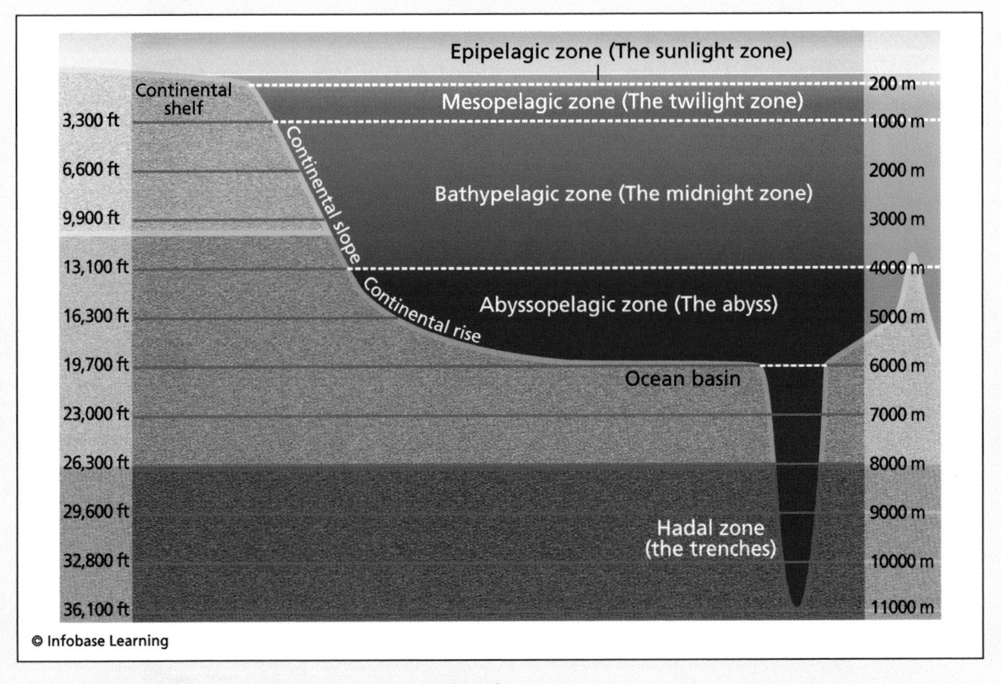



**Layers of the ocean** ***(Modeled after NOAA)***

*(continued)*

the sunlight zone, because most of the visible sunlight resides here. As sunlight penetrates the epipelagic zone, it supports photosynthesis by tiny marine creatures called phytoplankton. Bluefin tuna, sharks, swordfish, and other fast-swimming marine predators reside in this part of the open sea.

Solar heating is responsible for the wide change in temperature that occurs in the epipelagic zone. The amount of arriving sunlight depends upon both latitude and season. Consequently, sea surface temperatures can vary from a high of 97°F (36°C) in the Persian Gulf to just 28°F (–2.2°C) in the Arctic Ocean near the North Pole. (Note that salt water freezes at a lower temperature than freshwater.) Wind action mixes the water in this layer and causes a vertical distribution of incoming solar energy.

Immediately below the epipelagic zone lies the mesopelagic zone, which extends from 660 feet (200m) down to a depth of some 3,300 feet (1,006 m). A phenomenon called the thermocline characterizes the mesopelagic zone. The thermocline represents the transition layer between the wind-mixed surface water of the epipelagic zone and much deeper ocean water. In the thermocline, the water temperature decreases rapidly with increasing depth. The depth and magnitude of the thermocline varies from season to season and year to year. The phenomenon is typically most pronounced in the Tropics and essentially negligible in the polar regions during winter. Scientists sometimes refer to the

the salty oceans annually. While only a small, variable constituent of Earth's atmosphere, water vapor (gaseous water) plays a significant role in shaping the planet's overall weather and climate.

The oceans regulate Earth's climate in three major ways. First, currents help distribute incoming solar radiation from the warm equatorial regions of the planet to its colder, higher-latitude regions. Evaporation of warm ocean water as part of the hydrologic cycle is the second major way that the oceans influence Earth's climate. Finally, the oceans help keep Earth's climate habitable by releasing oxygen through phytoplankton photosynthesis and by serving as a huge repository for carbon dioxide ($CO_2$) removed from the atmosphere. Carbon dioxide is an important greenhouse gas. Scientists are now concerned that if too much carbon dioxide gets dissolved in water (fresh or salty), the water will become mesopelagic zone as the *twilight zone,* because any sunlight that penetrates into the zone is very faint. Due to a lack of sunlight, marine creatures that reside in this layer sometimes exhibit bioluminescence. Other marine inhabitants of this zone have developed very large eyes so they can more easily detect any prey appearing above them silhouetted against the dim light.

Ocean depths ranging from 3,300 feet (1,006 m) to 13,100 feet (3,993 m) characterize the bathypelagic (or bathyal) zone. The water temperature in this zone never varies far from a chilly 39°F (4°C). Scientists also refer to this zone as the *midnight zone* because it is a region of constant darkness. The only light available here comes from the bioluminescence of the marine creatures that inhabit this layer of the ocean.

The abyssopelagic (or abyssal) zone extends from 13,100 feet (3,993) down to the ocean floor, here nominally represented as 19,700 feet (6,005 m). This is the icy cold (near freezing temperature), pitch-black bottom layer of the ocean. Despite the frigid temperatures and crushing pressures, marine biologists have detected several intriguing creatures that inhabit the ocean basin.

Trenches form the deepest parts of the ocean. The hadalpelagic zone extends from the floor of the ocean basin, nominally a depth of 19,700 feet (6,005 m), to the bottom of the deepest region, a depth of 35,797 feet (10,911 m) in the Mariana Trench in the Pacific Ocean. Total darkness, extreme pressures, and near freezing water temperatures characterize the hadalpelagic zone.

more acidic, and the increase in acidity could adversely influence aquatic life, including the production of oxygen by phytoplankton.

The oceans provide a habitat for more than 250,000 known species of marine animals and plants. Fish and shellfish serve as important food sources for human beings and other land-dwelling creatures. While the oceans are vast and generally bountiful, the stress of overfishing by modern fishing fleets threatens to deplete many important fish stocks. Without effective international controls and agreements, overfishing of such species as cod, haddock, tuna, and swordfish may lead to commercial extinction. Environmental scientists define commercial extinction as the point when the population of a particular marine species decreases to a level at which it is no longer commercially viable to fish for that species. It takes decades for a severely depleted fish species to recover—if ever.

## OCEAN CONVEYOR BELT

The world ocean is not a still body of water. Ocean waves and currents are typically driven by surface winds and distribute energy around the globe, thereby exerting a huge impact on climate. In addition to the influence of the atmospheric winds, scientists recognize that there is constant motion deep within the world ocean due to a globally extensive ocean conveyor belt. The accompanying figure is a very simplified representation of this important oceanic phenomenon. It involves density driven thermohaline currents and is based on the physical principle that cold salty water (blue in the figure) is dense and sinks to the bottom of the ocean, while warm salty water (red) is less dense and rises to the surface. Thermohaline circulation takes place deep within the ocean and acts as a huge conveyor belt that redistributes large quantities of incoming solar radiation around the globe.

For ease of discussion, scientists typically assume the ocean conveyor belt starts in the Norwegian Sea portion of the North Atlantic Ocean. Warm water from the tropical regions of the Atlantic is transported to the Norwegian Sea by the Gulf Stream. The warm water of the Gulf Stream provides heat to the colder atmosphere found in the northern latitudes, especially during winter. This fortuitous release of thermal energy significantly moderates the climate of northwest Europe.

As the warm surface water releases heat to the atmosphere, it gets colder and experiences an increase in density, causing it to sink toward the ocean floor. As the colder water sinks, it moves south and makes room for newly arriving warm water. The colder bottom water flows south past the equator and travels all the way south to Antarctica. After the colder water passes through the Southern Ocean, it gradually rises to the surface and is warmed by the Sun as it travels through the Pacific Ocean and/or the Indian Ocean. The warm, less dense water then completes its cyclical journey by flowing back into the Atlantic Ocean. Scientists estimate that water takes about 1,000 years to make a complete loop around the globally extensive ocean conveyor belt.

Disruption of the ocean conveyor belt would have serious impacts on Earth's climate, especially that of northern Europe. Scientists are considering

## PACIFIC OCEAN

The Pacific Ocean is the largest of the Earth's five oceans. With a total surface area of 60.1 million square miles (155.557 million km$^2$), the Pacific Ocean covers about 28 percent of Earth's total surface. This huge body of

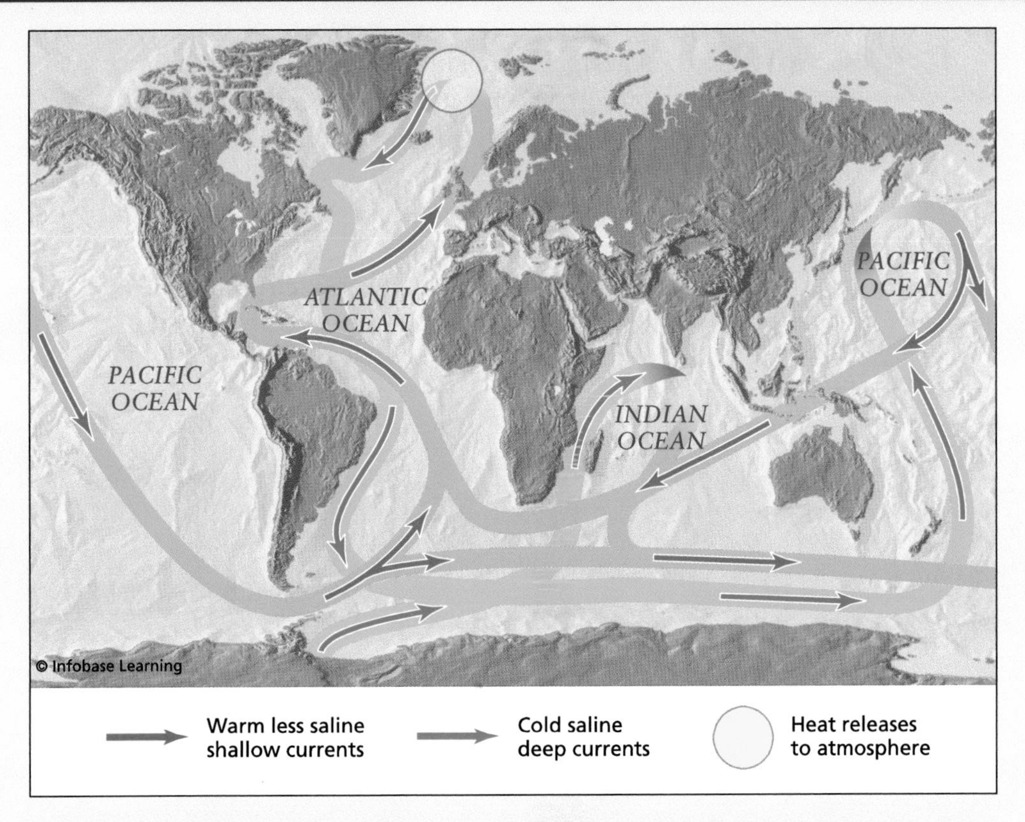


**Simplified depiction of the ocean conveyor belt—a global phenomenon involving density-driven thermohaline currents. Red represents warm (less dense) surface water; blue represents cold (more dense) deep water.** ***(Modeled after NOAA)***

the environmental consequences should sufficient quantities of freshwater from melting Arctic ice due to global warming mix with the salty surface water of the North Atlantic. It is possible that the salinity of the ocean's surface water could decline so severely that the formation of deep-water cold masses could no longer occur. This would then weaken or even completely shut down the ocean conveyor belt, making northern Europe much colder.

water is situated between the Southern Ocean, Asia, Australia, and the Western Hemisphere. As discussed here, the Pacific Ocean includes the Bali Sea, the Bering Sea, the Bering Strait, the Coral Sea, the East China Sea, the Gulf of Alaska, the Gulf of Tonkin, the Philippine Sea, the Sea of

Japan, the Sea of Okhotsk, the South China Sea, the Tasman Sea, and other tributary water bodies. In 2000, members of the International Hydrographic Organization decided to delimit a fifth global ocean, calling it the Southern Ocean. Their actions removed (on maps and charts) the portion of the Pacific Ocean lying south of the 60° south latitude line.

Earth's equator provides a convenient demarcation line between the North Pacific Ocean and the South Pacific Ocean. Surface currents in the North Pacific Ocean are dominated by a clockwise warm-water gyre. Oceanographers define a gyre as a broad, circular system of cur-

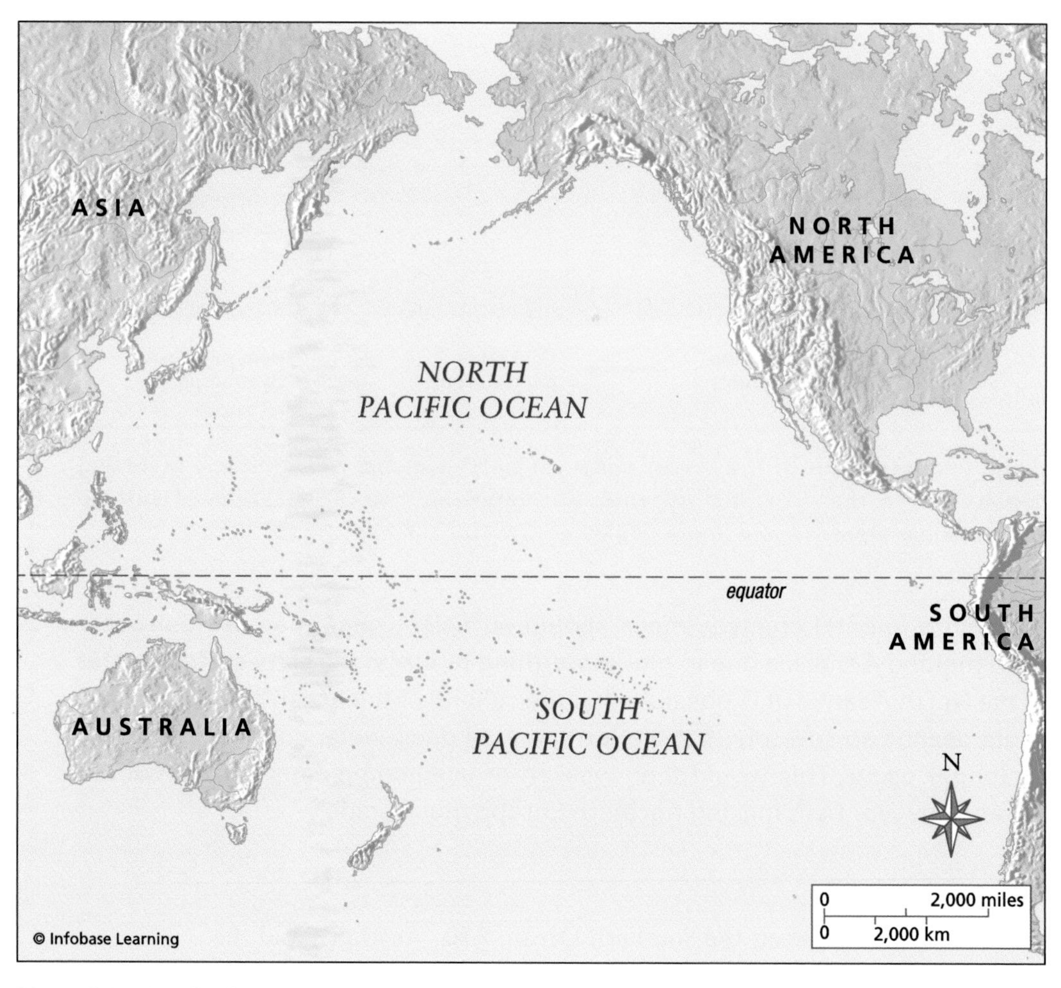


Map of the Pacific Ocean

rents. Surface currents in the South Pacific Ocean are dominated by a counterclockwise cool-water gyre. During the winter (especially the months of January and February) in the North Pacific Ocean, sea ice forms in the Bering Sea and the Sea of Okhotsk. During winter (especially the months of July and August) in the South Pacific, sea ice from Antarctica (Southern Ocean) reaches its northernmost extent in October. The ocean floor in the eastern portion of the Pacific Ocean is dominated by a feature scientists call the Eastern Rise, while the western portion is dissected by deep trenches, including the Mariana Trench, which is the world's deepest. The deepest point in the Pacific Ocean is the Challenger Deep in the Mariana Trench, which lies 35,840 feet (10,924 m) below sea level.

The Pacific Ocean is surrounded by a zone of violent volcanic and earthquake phenomena that scientists call the Pacific Ring of Fire. Tropical cyclones (hurricanes) can originate at different times of the year, striking Central America and Mexico from June to October and East and Southeast Asia from July to October. The cyclical El Niño/La Niña phenomena occur in the equatorial regions and can influence weather in the Western Hemisphere as well as regions of the western Pacific Ocean. El Niño ("the little boy") is characterized by unusually warm temperatures and La Niña ("the little girl") by unusually cool temperatures in the equatorial Pacific.

El Niño is an oscillation of the ocean-atmosphere system in the tropical Pacific that has important consequences for weather around the globe. El Niño's environmental consequences include increased rainfall across the southern tier of the United States and drought in the western Pacific. In normal, non–El Niño conditions, the trade winds blow toward the west across the tropical Pacific Ocean. These winds pile up warm surface water in the western Pacific, so that the sea surface is about 1.64 feet (0.5 m) higher at Indonesia than at Ecuador. During El Niño, the trade winds relax in the central and western Pacific, leading to a depression of the thermocline in the eastern Pacific and an elevation of the thermocline in the west. Oceanographers define the *thermocline* as the zone in which ocean water temperature decreases from warm surface values to colder, deepwater values.

La Niña is characterized by unusually cold ocean temperatures in the equatorial regions of the Pacific compared to El Niño, which is characterized by unusually warm ocean temperatures in the same regions. Scientists

## HAWAIIAN MONK SEAL

Scientists consider the Hawaiian monk seal *(Monachus schauinslandi)* one of the world's rarest marine mammals. At one time during the 20th century there were three members of the monk seal species: the Hawaiian monk seal, the Mediterranean monk seal, and the Caribbean monk seal. Unfortunately, relentless slaughter by humans through the mid-20th century drove the Caribbean monk seal to extinction, with the last confirmed sighting in the wild in 1952.

Monk seals live in warm, subtropical waters and spend about 67 percent of their lives at sea. Their habitat includes waters surrounding atolls, islands, and areas farther offshore on reefs and submerged ocean banks. When not swimming in the sea, monk seals like to relax and sleep on sandy, protected beaches. They also breed and raise the young seal pups in such sheltered environments.

Hawaiian monk seals were hunted to the brink of extinction in the 19th century. Today, fewer than 1,200 individuals remain. Entanglement in marine debris such as carelessly discarded plastic products, disease outbreaks, and capture in commercial fishing nets are some of the major threats facing this endangered seal. Natural predators include sharks, which prey on monk seals while they are at sea foraging for food. All Hawaiian monk seals live within the territorial waters of the United States, with the majority of the seals found in six main breeding subpopulations in the northwestern Hawaiian Islands.

Male monk seals reach 7.1 feet (2.16 m) in length and 375 pounds-mass (170 kg) in mass. Female monk seals are slightly larger and reach a length of 7.5 feet

suggest that the global climate impacts of La Niña tend to be the opposite of those caused by El Niño. For example, during a La Niña year, winter temperatures tend to be warmer than normal in the southeastern part of United States and cooler than normal in the northwestern part of the country.

The Pacific Ocean is a major contributor to the world economy and exerts particular influence on those nations that border its water directly. The vast ocean accommodates low-cost sea transportation between nations in the Eastern and Western Hemispheres. The ocean's extensive fishing grounds provide more than 60 percent of the world's fish catch. Combating oil pollution, avoiding overfishing, and protecting endangered marine species are some of the major environmental issues facing scientists this century.

(2.3 m) and a mass of 450 pounds-mass (205 kg). Monk seals have a life expectancy of 25 to 30 years. Scientists classify monk seals as benthic foragers. Adult seals commonly feed at night, hunting for food outside the immediate shoreline areas in waters that are some 60 to 300 feet (18–91 m) deep. In contrast, juvenile seals hunt near the shoreline in the daytime, searching for marine species that hide in the sand or under rocks.

**An endangered Hawaiian monk seal swims in the Pacific Ocean near Laysan Island, Hawaii.** ***(NOAA)***

## ATLANTIC OCEAN

With a total surface area of 29.65 million square miles ($76.762 \times 10^6$ km$^2$), the Atlantic Ocean is the planet's second-largest ocean. It is located between Africa, Europe, the Southern Ocean, and the Western Hemisphere. As discussed here, the Atlantic Ocean includes the Baltic Sea, the Black Sea, the Caribbean Sea, the Davis Strait, the Denmark Strait, a portion of the Drake Passage, the Gulf of Mexico, the Labrador Sea, the Mediterranean Sea, the North Sea, the Norwegian Sea, almost all of the Scotia Sea, and other tributary bodies of water. In 2000, members of the International Hydrographic Organization decided to delimit a fifth world ocean, the Southern Ocean, and removed the portion of the Atlantic Ocean south

*(continues on page 78)*

## NORTH ATLANTIC RIGHT WHALE

With between 300 and 400 animals remaining, the North Atlantic right whale is one of the world's most endangered marine mammals. Hunted to the brink of extinction by Yankee (American) whalers, stringent laws and agreements such as the Marine Mammal Protection Act (MMPA) and the Endangered Species Act (ESA) are now attempting to help the few remaining right whales recover as a viable marine species.

The right whale received its name because it was quite literally the "right whale" to hunt. The gentle marine mammal moves slowly through the water and floats after being killed by a whaler's harpoon. Because of these characteristics, Yankee whalers from New England aggressively hunted the species to near extinction in the 19th century. Despite being protected by federal law for the past 50 years, the right whale still has not recovered from the relentless slaughter that severely depleted its population.

North Atlantic right whales are found year round from Cape Cod (Massachusetts) to Nova Scotia (Canada). The right whales use this area of the North Atlantic to feed and mate. Then, in the fall, pregnant female right whales travel down the eastern coast of the United States to have their calves (baby whales) in the warm waters off the southeastern U.S. Atlantic Coast. Scientists regard the Atlantic coast from about Cape Canaveral, Florida, to Brunswick, Georgia, as a critical habitat for right whales. Here, after giving birth in the spring, the female right whales nurse their young calves for about 12 months, after which both mother and calf begin the long journey back north.

The right whale migration route hugs the Atlantic coast of the United States. Unfortunately, this route takes the endangered right whales through very heavily traveled coastal waters. The whales spend a great deal of time on or near the surface and are generally oblivious to any of the dangers imposed by human activities. To aggravate the situation, right whales are very difficult to see because of their dark skin and lack of dorsal fin. Consequently, strict federal laws and regulations have been imposed to protect the right whales during their migratory journeys up and down the coast. It is against federal law for any type of watercraft, aircraft, swimmer, or diver to approach or remain within 500 yards (457 m) of a North Atlantic right whale. Violators face severe criminal and civil penalties. In order to protect each remaining right whale, the U.S. Coast Guard helps enforce such marine mammal protection laws.

Scientists report that about one-third of all known North Atlantic right whale mortalities are the result of vessel collisions or entanglement in fixed fishing

A North Atlantic right whale—the world's most endangered whale—lunges out of the ocean at NOAA's Stellwagen Bank National Marine Sanctuary off Scituate, Massachusetts. *(NOAA)*

gear. Collision with a marine vessel as small as 43 feet (13.1 m) in length can kill a right whale. The loss of just one or two reproductive-age female right whales per year can lead to the extinction of the species.

The right whale can reach a maximum length of 60 feet (18.3 m) and a maximum mass of 100 tons (91 metric tonnes). As filter feeders, they typically search for dense patches of zooplankton called copepods. They may also eat small fish near the ocean floor. Scientists estimate that right whales have a natural life expectancy of about 100 years. Illegal hunting by rogue whaling ships is also a danger to survival. Agencies of the U.S. government, such as NOAA's Fisheries Service, diligently try to protect all right whales found in territorial waters. Unfortunately, outside American coastal waters, despite international agreements to protect all endangered whales, the southern right whale and the North Pacific right whale still fall victim to illegal slaughter by rogue whaling ships from other nations.

*(continues)*

*(continued)*

The International Whaling Commission (IWC) was established in 1946 to help conserve whale stocks and to regulate catches of endangered whales by commercial whaling operations. Since 1986, the IWC has supported a global moratorium on commercial whaling, nominally to allow depleted/endangered species time to rebuild a viable percentage of their previous populations. Modern commercial whaling is a very inflammatory international issue. Pro-whaling countries such as Japan, Iceland, and the Russian Federation support an increase in whaling activities; antiwhaling countries such as the United States and environmental groups want to extend and expand the ban on commercial whaling activities.

Despite the establishment of IWC-sponsored whale sanctuaries in the Antarctic region, the Indian Ocean, and most recently the Southern Ocean, rogue whaling activities still occur in these supposedly protected ocean regions. Furthermore, in response to the IWC ban on commercial whaling, some IWC member nations such as Japan and Iceland have instituted major research programs that include the capture and killing of whales under permitted "scientific sampling" activities. The IWC also allows subsistence whaling of various species by aboriginal peoples.

*(continued from page 75)*

of 60° south latitude. The equator divides the North Atlantic and the South Atlantic.

The Atlantic Ocean contains some of the world's most heavily trafficked sea routes between and within the Eastern and Western Hemispheres. Major choke points and strategic passageways include the Dardanelles (Turkey), Strait of Gibraltar (Spain-Morocco), Panama Canal, Suez Canal, Oresund (Denmark-Sweden), Kiel Canal (Germany), Saint Lawrence Seaway (United States-Canada), Mona Passage, Straits of Florida, and Strait of Dover. Crude oil and natural gas energy resources are found in the Gulf of Mexico, Caribbean Sea, and North Sea.

The North Atlantic has a clockwise warm-water gyre, while the South Atlantic has a counterclockwise warm-water gyre. A rugged north-south centerline called the Mid-Atlantic Ridge dominates the ocean floor of the entire Atlantic basin. The deepest point in the Atlantic Ocean is the Milwaukee Deep, located in the Puerto Rico Trench at a depth of 28,232 feet (8,605 m) below sea level.

The environmental hazards found in the Atlantic Ocean include icebergs and hurricanes (tropical cyclones). A threat to ships traveling across the North Atlantic, icebergs have been spotted as far south as Bermuda and the Madeira Islands, an autonomous region of Portugal in the mid-Atlantic. One of the greatest maritime disasters to occur during peacetime started late on the evening of April 14, 1912, when the British luxury passenger ship RMS *Titanic* struck an iceberg during its maiden voyage across the North Atlantic. The doomed ship disappeared below the surface of the frigid waters early on the morning of April 15. The iceberg collision claimed the lives of more than 1,500 passengers and crew. Hurricanes typically occur from May to November in the North Atlantic. These powerful storms develop off the coast of Africa near Cape Verde, an island nation off the western coast of Africa, and travel westward across the ocean toward the Caribbean Sea. Hurricanes can also develop in the Gulf of Mexico or in the Caribbean Sea. Oil spills, municipal sewage pollution, the decline of fish populations due to overfishing, and the need to more effectively protect endangered marine mammals are some of the major environmental issues facing the nations that border the Atlantic Ocean.

## INDIAN OCEAN

With a surface area of 26.48 million square miles ($68.556 \times 10^6$ km$^2$), the Indian Ocean is Earth's third-largest ocean. As discussed here, the Indian Ocean includes the Andaman Sea, the Arabian Sea, the Bay of Bengal, the Flores Sea, the Great Australian Bight (large bay), the Gulf of Aden, the Gulf of Oman, the Java Sea, the Mozambique Channel, the Persian Gulf, the Red Sea, the Savu Sea, the Strait of Malacca, the Timor Sea, and other tributary bodies of water. Four strategic waterways—the Suez Canal (Egypt), Bab el Mandeb (Djibouti-Yemen), Strait of Hormuz (Oman-Iran), and Strait of Malacca (Malaysia-Indonesia)—provide access to the Indian Ocean. By international agreement in 2000, scientists assigned the portion of the Indian Ocean south of 60° south latitude to the Southern Ocean.

The Indian Ocean provides major sea routes connecting the Middle East, Africa, and East Asia with Europe and the Western Hemisphere. This ocean region carries especially heavy shipping traffic involving petroleum and petroleum products from the oilfields of the Persian Gulf and Indonesia. Consequently, oil pollution in the Persian Gulf, Arabian Sea, and Red Sea is a major environmental issue. Overfishing is another environmental issue. The fish resources of this ocean are of major importance to the bordering countries such as India, Pakistan, and Thailand for domestic consumption and export. Commercial fishing fleets from Japan, South Korea, Taiwan, and Russia also exploit the marine resources of the Indian Ocean. Effective protection of endangered marine mammals remains another pressing environmental issue. An occasional iceberg in the southern portions of the Indian Ocean represents a natural hazard to shipping. Finally, despite internationally sponsored countermeasures, commercial shipping and pleasure craft traveling the waters of the Indian Ocean remain at high risk to modern acts of piracy, especially in the Gulf of Aden, along the east coast of Africa, in the Bay of Bengal, and in the Strait of Malacca.

The Indian Ocean is dominated by a counterclockwise gyre in the southern region and a unique reversal of surface currents in the northern region. Hot, rising summer air over southwest Asia causes low atmospheric pressure that results in the southwest monsoon (June to October) and southwest-to-northeast winds and currents. Cold air that descends over northern Asia in winter (December to April) produces the northeast monsoon and northeast-to-southwest winds and currents. Tropical cyclones (hurricanes) occur in the northern portions of Indian Ocean from about May through November and in the southern portions of the Indian Ocean from January to February.

The Mid–Indian Ocean Ridge dominates the ocean floor. Scientists have subdivided this physical feature into three regions: the Southeast Indian Ocean Ridge, the Southwest Indian Ocean Ridge, and the Ninetyeast Ridge. The deepest point in the Indian Ocean is the Java (Sunda) Trench, with a maximum depth of about 25,344 feet (7,725 m) below sea level.

## SOUTHERN OCEAN

The Southern Ocean is the body of water between 60° south latitude and the continent of Antarctica. This ocean has a total surface area of 7.85 million square miles ($20.327 \times 10^6$ km$^2$) and includes the Amundsen Sea, the Bellingshausen Sea, part of the Drake Passage, the Ross Sea, a small portion of the Scotia Sea, the Weddell Sea, and other tributary water bodies. The Drake Passage lies between South America and Antarctica.

Oceanographic research performed in the late 20th century indicated that an ocean current, the Antarctic Circumpolar Current (ACC), flows from west to east around Antarctica. This current plays a crucial role in global ocean circulation. The Antarctic Convergence (or simply Convergence) is the region where the cold waters of the ACC mingle with the warmer waters of the Atlantic, Pacific, and Indian Oceans found farther north. Scientists recognize that the Convergence encompasses a discrete body of water and a unique ecological region. Specifically, the Convergence concentrates nutrients that promote marine plant life. This, in turn, supports a greater abundance of marine animal life, including krill, Chilean sea bass (Patagonian toothfish), squid, seals, and whales. In spring 2000, the International Hydrographic Organization (IHO) decided to delimit the waters within the Convergence and named this environmentally distinct body of water the Southern Ocean. By defining the Southern Ocean as extending from 60° south latitude to the coast of the Antarctic Continent, IHO members integrated the southern waters of the Atlantic, Pacific and Indian Oceans.

Although the Convergence fluctuates annually, it generally occurs in the middle of the ACC and therefore serves as a natural dividing line between the colder polar surface waters found to the south and the warmer ocean waters found to the north. The Southern Ocean is subject to all the international agreements regarding the world's oceans. It is also subject to numerous international agreements regarding Antarctica, including the international prohibition of commercial whaling and the Convention on the Conservation of Antarctic Marine Living Resources. Despite these international restrictions, rogue whaling and commercial fishing operations do occur in the environmentally protected waters of the Southern Ocean. The ocean's remote location and frequent harsh weather conditions make enforcement of the protective international regulations by an appropriate authority very difficult, especially since various treaties specify that Antarctica and its resources are not subject to national sovereignty.

Sea surface temperatures vary from about 50°F (10°C) to 28.4°F (–2°C). Cyclonic storms travel eastward around the continent and are intense because of the temperature contrast between ice and the open ocean. From 40° south latitude to the Antarctic Circle (approximately 66° south latitude), the open ocean contains the strongest average winds found anywhere on Earth. During the winter, the ocean freezes outward from the Antarctic coast to about 65° south latitude in the Pacific sector and to about 55° south latitude in the Atlantic sector.

With an average depth ranging from 13,123 to 16,404 feet (4,000 to 5,000 m), the Southern Ocean is a deep body of water containing only a few shallow areas. The Antarctic continental shelf is generally narrow and unusually deep. With a depth of 23,737 feet (7,235 m), the southern end of the South Sandwich Trench is the lowest point in the Southern Ocean. The Antarctic Circumpolar Current moves perpetually eastward and is the world's largest ocean current (13,050 mi [21,000 km] in length). Scientists estimate that this current transports 4,590 million cubic feet of water per second ($130 \times 10^6$ $m^3/s$), a flow rate that is 100 times greater than the flow of all the world's rivers combined.

## ARCTIC OCEAN

With an estimated total area of 5.43 million square miles ($14.056 \times 10^6$ $km^2$), the Arctic Ocean is the smallest of the planet's five oceans. As discussed here, the Arctic Ocean includes Baffin Bay, the Barents Sea, the Beaufort Sea, the Chukchi Sea, the East Siberian Sea, the Greenland Sea, Hudson Bay, the Kara Sea, the Laptev Sea, and the Northwest Passage. The Arctic Ocean is the body of water quite literally at the top of the world, separating Europe, Asia, and North America.

Surface ships that attempt to transit the Arctic Ocean are often impaired by sea ice. The Chukchi Sea is a major choke point. Named for the indigenous Chukchi people of far eastern Siberia, the Chukchi Sea is an extension of the Arctic Ocean that separates Alaska (United States) from Siberia (Russia). Frozen for most of the year, portions of this sea are typically navigable during the months August through October, providing nautical access from the Arctic Ocean to the North Pacific Ocean via the Bering Strait. The Northwest Passage (adjacent to the United States and Canada) and the Northeast Passage (or Northern Sea Route, adjacent to Norway and Russia) are two potentially important seasonal waterways. For example, the Northeast Passage is the shortest marine route between the extremes of eastern and western Russia.

The Arctic Ocean has a polar climate characterized by persistent cold and relatively narrow yearly temperature ranges. During the arctic summer, the region experiences continuous daylight, the weather is foggy and damp, and an occasional weak cyclone occurs accompanied by snow or rain. During the arctic winter, the region is shrouded in continuous darkness, and the weather is cold and stable. The clear night skies often dazzle with the brilliant colors of the northern lights, also known as the aurora borealis.

A perennial, drifting polar ice pack with an average thickness of about 9.84 feet (3 m) covers the central portion of the Arctic Ocean. Open seas surround the edges of the pack ice during the summer months. During winter, the pack ice more than doubles its surface area and typically extends to the encircling landmasses in Europe, Asia, and North America. Scientists are now investigating the rapid melting and disappearance of the polar ice cap, probably a result of changes in global climate patterns. The observed decline of the polar ice cap causes environmental scientists to express alarm about the disruption of the fragile arctic ecological system and the threat to such native animal species as the polar bear *(Ursus maritimus).*

The floor of the Arctic Ocean is about 50 percent continental shelf, with the remainder consisting of a central basin. Three submarine ridges, Alpha Cordillera, Nansen Cordillera, and Lomonosov Ridge, interrupt the topography of the ocean basin. At a depth of 15,305 feet (4,665 m), the

Several curious polar bears approach the bow section of the American nuclear submarine USS *Honolulu* (SSN 718) after the ship surfaced in the Arctic Ocean some 280 miles (450 km) from the North Pole. As part of this mission in 2003, the submarine's crew collected environmental data for use by researchers in American and Canadian universities. *(DOD/USN)*

Fram Basin is the deepest known part of the Arctic Ocean. On March 17, 1959, the American nuclear submarine USS *Skate* (SSN-578) navigated through the Arctic Ocean under the pack ice and became the second submarine to reach the North Pole and the first vessel to break through the ice and surface there. The previous year, on August 3, 1958, the world's first nuclear submarine, the USS *Nautilus* (SSN-571) had crossed the North Pole by traveling through the Arctic Ocean under the pack ice. Since these historic polar voyages in the middle part of the 20th century, many other American nuclear submarines have navigated through and under the pack ice of the Arctic Ocean. These visits to the Arctic Ocean often involve the collection of important scientific data.

Starting in the 16th century, European explorers began searching for the Northwest Passage, a much-desired water route from the Atlantic Ocean across the northern portions of Canada and Alaska into the Pacific Ocean. It was not until 1969 that a commercial ship was able to complete this journey. In that year, a specially reinforced icebreaking supertanker, the SS *Manhattan*, started its historic journey in the Atlantic Ocean, traveled through ice-choked Canadian waters, and then arrived at Prudhoe Bay, Alaska. There it loaded a symbolic barrel of oil onboard and journeyed back across the Northwest Passage to the east coast of the United States. Although successful, this historic journey proved very difficult, so oil companies soon abandoned the idea of using icebreaking supertankers for sea transport of Alaskan oil by means of the Northwest Passage. The Trans-Alaska Pipeline System became the favored approach. However, the record recent summer melting of the Arctic Ocean's icepack has again stimulated interest in the regular use of maritime shipping lanes across both the top of North America via the Northwest Passage and across the top of Eurasia via the Northeast Passage. This renewed consideration of using the Arctic Ocean for commercial marine transit has raised political issues (for example, the legality of international shipping routes through Canada's polar archipelago) and environmental issues (such as the possible disruption of fragile polar ecosystems).

CHAPTER 6

# Water—Its Power and Applications

**Water is the key to life on Earth. This chapter discusses how people** have used and controlled water throughout recorded history. Past human civilizations grew and prospered when their people had access to abundant quantities of water for drinking, irrigation, and sanitation. The ability to travel across bodies of water with an evolving family of boats and ships opened up the globe to exploration, commerce, and communication. Water also provided an important early form of power, hydropower, to grind grain. Today, modern engineers have constructed enormous dams, elegant canals, and complex water supply systems with which to provide electric power, nautical transport, dependable water supplies, and efficient sanitary waste removal for millions of people.

## WATER—THE MIGHTY TRANSPORTER

Anyone who has viewed a waterfall, large or small, has witnessed the power of flowing water. Throughout Earth's history, the relentless natural flow of water has sculpted and shaped the surface of the planet.

Rocks at or near Earth's surface can experience weathering and erosion. The continual and relentless action of environmental forces in the atmosphere, such as wind, and the hydrosphere, such as flowing water, slowly disintegrate and decompose such rocks. Rainwater may seep into cracks in a rock and then freeze. As ice forms and expands in a crack, it

An aerial view of Victoria Falls on the Zambezi River in southern Africa, as seen from the eastern (Zambian) side. The famous waterfall is about one mile (1.6 km) wide and has a maximum drop of 420 feet (128 m); named in honor of Queen Victoria (1837–1901) of Great Britain in 1855 by the Scottish explorer David Livingstone (1813–73). *(CIA)*

can exert tremendous forces on the rock, often with sufficient strength to break it. Geologists refer to these natural erosion processes as *mechanical weathering.* There are many environmental actions that contribute to converting rocks into smaller sediments as part of the overall rock cycle.

Geologists describe the breakdown of surface rocks into smaller and smaller fragments using the following descriptive order, (from larger to smaller fragments): boulders, cobbles, gravel, sand, silt, and microscopic clay-sized particles. The finer the particle, the easier it is for natural forces to move it. Sediments are transported by water or wind erosion; rock fragments experience the dissolution and transport of their soluble chemical components by surface water and groundwater.

As part of the rock cycle, rock fragments are transported and deposited in a variety of sedimentary environments. On land, such sedimentary environments include swamps, stream flood plains, dunes, and desert basins. Along coastal regions, sediments can accumulate in the deltas at

## GRAND CANYON NATIONAL PARK

Grand Canyon National Park is a World Heritage Site that lies on the Colorado Plateau in northwestern Arizona. The Grand Canyon is an inspiring landscape that is 277 river miles (446 km) long, up to 18 miles (29 km) wide, and a mile (1.6 km) deep. Over time, drainage systems have cut deeply through the rock and formed numerous steep-walled canyons. Scientists consider the Grand Canyon one of the finest examples of arid land erosion. The exposed geological strata, layer upon layer from the basement Vishnu schist to the capping Kaibab limestone, rise more than one mile (1.6 km) above the Colorado River and represent one of the most complete records of geologic history that can be seen anywhere in the world. Geologists estimate that the rock formations, such as schist and gneiss, found at the bottom of the Grand Canyon date back 1.8 billion years. Between 70 and 30 million years ago, the relatively high and flat Colorado Plateau was formed due to the action of plate tectonics, which lifted up the entire region. Then, starting between 5 and 6 million years ago, the Colorado River began to carve its way downward. As the Colorado River deepened the magnificent canyon, water erosion by tributary streams widened it. This relentless natural sculpting by flowing water and wind continues to deepen and widen the Grand Canyon.

Spectacular view of the Colorado River from the Toroweap Overlook at the western end of Grand Canyon National Park *(NPS)*

the mouths of rivers, in lagoons, on beaches, and on barrier islands. The great majority of sediments ultimately come to rest in the world's oceans. The sediments can accumulate in massive deposits that form continental shelves or else continue to the deep ocean basins beyond.

As sediments accumulate over geologic time scales, the weight of the overlying materials causes compaction of the materials in the lower sedimentary layers. In addition to compaction, sedimentary materials can experience cementation, the process in which previously dissolved and transported minerals precipitate in pore spaces and join (cement) the grains of sediment together. Geologists refer to the combined effects of compaction and cementation as the process of lithification. This process results in the formation of sedimentary rocks, such as sandstone, chalk, and limestone. Over time, sedimentary rocks formed near Earth's surface typically undergo exposure to weathering and erosion as part of the overall rock cycle. The Grand Canyon in northern Arizona is a spectacular example of the sculpting power of flowing water.

As civilizations arose, people learned how to use the flow of water in a variety of clever and important ways. Ancient peoples learned how to channel water to satisfy thirst, to irrigate crops, and to wash away wastes. They also learned to travel on water in various vessels and how to capture the energy of flowing water to turn large stone wheels in order to grind grain more efficiently. Modern versions of these technical achievements form the framework of 21st-century life in developed portions of the globe.

## ANCIENT WATERWORKS

Anyone who travels to a very rural region within the United States could have the opportunity to hand-pump potable water from a well or else use an outdoor toilet facility, commonly called an outhouse. That experience usually creates a deeper appreciation of the personal conveniences that modern sanitary engineers, who understand hydraulics, microbiology, environmental technology, and related disciplines, have made possible. People often associate the availability of safe water for drinking, bathing, cooking, cleaning, and waste removal as one of the hallmarks of modern civilization. Many businesses and industries also depend on a stable supply of freshwater. Irrigation agriculture is just one example. Modern engineers often refer to the entire system of reservoirs, tanks, pipes, wells, pumps, and processing equipment, which operate together to deliver potable water to a municipality and remove runoff water and sanitary waste streams, as simply the waterworks.

Human-engineered waterworks are as old as the earliest cities of ancient civilizations. As more and more people came together to live in cities, the first urban planners and ancient engineers began to divert naturally flowing water into these settlements for drinking, bathing, and carrying away human wastes. Many of the pre-Roman civilizations around the Mediterranean Sea relied upon simple ditches, crude clay pipes, and leaky aqueducts to channel freshwater into an urban region and then to carry untreated waste away—typically back into a nearby stream or river. The trick, of course, was to take in freshwater upstream from the location where the raw sewage was being poured back into the stream or river.

Unfortunately, even in the technology-rich 21st century, the impoverished inhabitants of many highly populated third world countries have little control over where raw waste enters a natural body of flowing water or where they can bathe, wash clothes, or draw water for drinking and cooking. Such conditions can lead to the spread of a wide variety of diseases, including outbreaks of typhoid and cholera.

As master builders of the ancient world, the Romans were the first to construct some of the world's most elaborate public sewage systems. According to historic tradition, construction of the Cloaca Maxima (Latin for "Great Sewer") began in Rome about 600 B.C.E. This early sewer system removed human waste from the rapidly growing ancient city and carried the effluent into the nearby Tiber River.

Although the Romans were not the first to construct aqueducts, they built an enormous network of well-engineered water-channeling structures out of stone, brick, and mortar. Carefully surveyed and constructed, these aqueducts used gravity to transport water over great distances to Roman cities. One of the earliest was the Aqua Appia, which was built around 312 B.C.E. to carry water to Rome. A later aqueduct, called the Aqua Marcia, carried water to Rome from a natural source located about 57 miles (91 km) away. Careful Roman engineering and the innovative use of cement and concrete made an aqueduct's flow channel watertight. At the peak of the empire, a system of nine major aqueducts carried millions of gallons (liters) of water to Rome each day.

Roman engineers also built aqueducts throughout the empire. Between 312 B.C.E. and 455 C.E., they constructed more than 200 aqueducts to carry water from distant springs, lakes, and rivers to thriving urban centers in France, Germany, Spain, Portugal, North Africa, and Asia Minor. One of the major differences between the ancient Roman waterworks and modern water supply and sanitary sewer systems is that the Roman engineers did not generally use impoundment dams to create water reservoirs. They chose, instead, to use the aqueducts to divert water from natural springs, lakes, and free-flowing streams to urban centers.

Urban life in imperial Rome required an enormous quantity of water. Heated bathhouses served as a central social activity for most Roman citizens. These bathhouses often adjoined public toilet facilities. The latrines in the public toilets generally included marble seats and running water that continuously flushed human waste into a sanitary sewer system. Roman engineers used the aqueducts to supply the capital city with freshwater for drinking and for the public baths. The continuous flow of water also removed wastes from the city's sewer system complex.

The Cloaca Maxima sewer system served Rome until the Western Empire fell. As invading barbarians attacked Rome in the fifth century, they captured the aqueducts and cut off the flow of water into the city, thereby crippling the cleansing operation of the sewer system. For example, the Visigoth king Alaric I (ca. 370–410 C.E.) attacked and sacked the famous city on August 24, 410. Prior to the complete collapse of the Western Empire, nominally dated by historians as 476, Rome's sewer system was widely imitated throughout Roman-controlled Europe. Portions of the Cloaca Maxima, including its outfall into the Tiber River, can still be seen in the modern city of Rome.

## MODERN WATERWORKS

A constant supply of clean, safe drinking water is essential for the maintenance of every community large and small. The inhabitants of large cities often drink tap water that comes from surface water sources such as lakes, rivers, and reservoirs. Sometimes these sources are relatively close to the municipality; other times the water source is many miles (km) away and is delivered to the municipal waterworks by a system of pumps, large pipelines, and modern aqueducts.

In many rural areas in the United States and around the world, people are more likely to drink groundwater that has been pumped from a well. These wells tap natural underground reservoirs called aquifers. The quantity of water produced by a well depends on the characteristics of the rock, soil, and sand in the aquifer. Drinking water wells can be relatively shallow, with a depth of 50 feet (15 m) or less, or quite deep, perhaps 1,000 feet (305 m) or more.

In the United States, large-scale water supply systems typically rely on surface water sources, while smaller municipal water systems tend to use groundwater. Generally, an underground network of pipes delivers treated drinking water to the homes and businesses served by the municipal (or local) water system. In the United States (as in many developed nations), drinking water must meet required health standards when it leaves the treatment plant. Engineers and technicians also monitor the treated drinking water after it leaves the treatment plant and travels through the distribution system. Their job is to identify and quickly remedy any problems that can affect the quality or quantity of the water being delivered. Potential problems include water main breaks, undesirable pressure variations, the growth of harmful microscopic organisms, or even intentional acts of terrorism.

## TREATED DRINKING WATER USE

Scientists at the U.S. Environmental Protection Agency (EPA) estimate that the average American currently uses between 90 to 100 gallons (340 to 379 L) of water each day in the home. The typical American household uses approximately 107,000 gallons (405,000 L) annually. Most of this drinking water (that is, water that has been treated to drinking water standards) goes to flushing toilets; to washing clothes, dishes, and automobiles; and to watering lawns. In many parts of the country, some 50 to 70 percent of home water consumption involves the watering of lawns and gardens or the filling and maintenance of swimming pools. Unrepaired leaks in household plumbing waste about 14 percent of the water used by an average American household.

EPA scientists also report that Americans use much more water each day than individuals in other developed countries as well as individuals in undeveloped countries. The average European uses about 53 gallons (201 L) of water per day, while the average person in sub-Saharan Africa uses a mere three to five gallons (11 to 19 L) of water per day. By learning how to wisely use water at home and at work, American citizens not only will help protect and improve their environment but also will personally benefit from a reduction in monthly water utility bills.

EPA scientists point out that on an average day in the United States water utilities process some 34 billion gallons (129 billion L) of water. The source and quality of water determines the specific amount and type of treatment required to meet drinking water standards. Surface water systems such as lakes and reservoirs generally require more treatment than groundwater systems because they are exposed to contaminants in the atmosphere and runoff from rain and melting snow.

Municipal waterworks employ a variety of treatment processes to remove contaminants from drinking water. Water treatment engineers often arrange these processes in a sequence called a treatment train. The most commonly used processes include flocculation (coagulation) and sedimentation, filtration, and disinfection. Engineers at some plants also use ion exchange and adsorption processes in the treatment train. Water treatment plant personnel select the precise combination that is most appropriate to process the incoming source water.

During flocculation, dirt and suspended particles are removed from source water. Water treatment technicians add alum, iron salts, or synthetic organic polymers to the incoming water. These additives form sticky particles called *floc* that attract dirt particles. In sedimentation, gravity pulls the heavy flocculated particles down through the water and makes them settle on the bottom of the treatment tank. Treated water next moves from the sedimentation tank to the filtration process, where a variety of filters, including sand and charcoal, gradually remove most, if not all, of the smaller particles, including organic matter, microorganisms, and suspended clays and silts. The filtration process greatly clarifies the water and enhances the final treatment step, disinfection.

Sanitary engineers and government officials regard the disinfection of public drinking water as one of the major public health advances of the 20th century. Many municipal water systems use chlorine, chlorinates, or chlorine dioxides to kill any dangerous microorganisms that remain in water before they release that water into the distribution system. In the United States, the fluoridation of drinking water helps prevent tooth decay.

---

*(opposite)* The major components of a complex municipal sanitary sewer system and its response to urban wet weather flows. Sanitary sewer overflows (SSOs) and several typical water pollution sources are included. *(Modeled after EPA)*

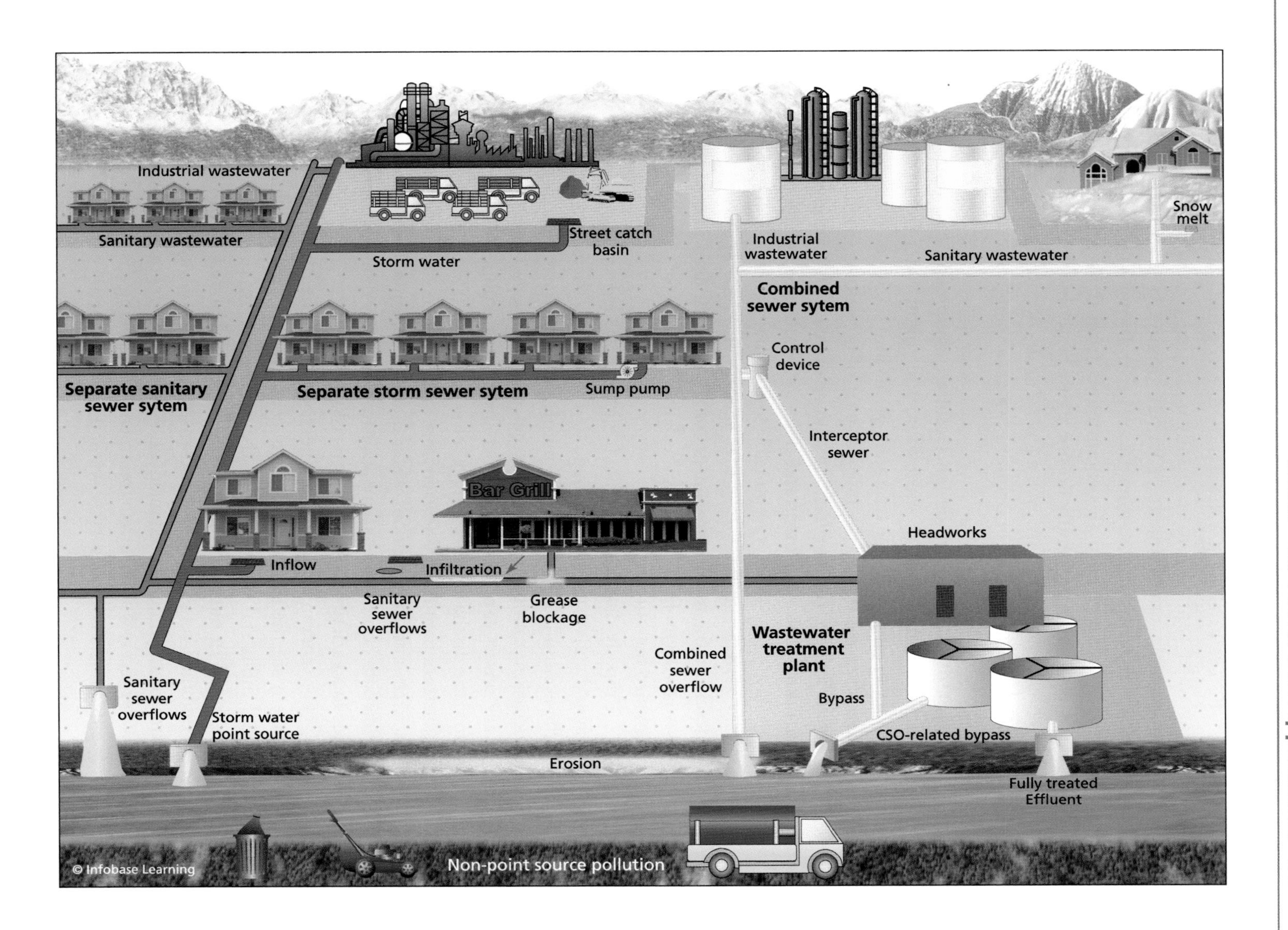
Industrial wastewater
Sanitary wastewater
Storm water
Street catch basin
Industrial wastewater
Sanitary wastewater
Snow melt
Combined sewer sytem
Control device
Separate sanitary sewer sytem
Separate storm sewer sytem
Sump pump
Interceptor sewer
Headworks
Inflow
Bar Grill
Infiltration
Sanitary sewer overflows
Grease blockage
Wastewater treatment plant
Combined sewer overflow
Bypass
Sanitary sewer overflows
Storm water point source
CSO-related bypass
Erosion
Fully treated Effluent
© Infobase Learning
Non-point source pollution

EPA safe drinking water standards recognize that all sources of drinking water contain some naturally occurring contaminants—in addition to possible contaminants of human origin. At low levels in drinking water, the natural contaminants are generally not harmful. It is extremely expensive to remove all contaminants from drinking water, and such costly activities generally do not significantly increase the protection of public health. There are even a few naturally occurring minerals that tend to improve the taste of drinking water, although the "taste of water" is very subjective.

Sanitary engineers design facilities, equipment, and fluid flow systems to handle sewage, wastewater, and runoff water from precipitation in urban areas. Sewage treatment plants and wastewater treatment plants are often considered an integral part of a complex, modern municipal water system. In some locations, however, one public utility provides safe drinking water to homes and industry, while another local government agency (or agencies) is responsible for the maintenance and repair of sanitary sewers and/or storm water sewers. The accompanying illustration describes the major components of a typical modern municipal sanitary sewer system. For simplicity, the delivery system for safe drinking water is implied but not depicted.

Properly designed, operated, and maintained sanitary sewer systems should collect all of the sewage that flows into them and then transport the accumulated sewage to a municipal wastewater treatment plant. On occasion, unintentional discharges of raw sewage from municipal sanitary sewers can occur in almost any system. Public health officials refer to these discharges as *sanitary sewer overflows* (SSOs). As depicted in the illustration, untreated sewage from an SSO can contaminate surface water sources and cause serious water quality problems. Raw sewage can also flow into homes and basements, cause property damage, and expose people to a variety of serious health risks.

For economic reasons, engineers designed the combined sewer system (again see the illustration) to collect rainwater runoff, domestic sewage, and industrial wastewater in the same pipe. This type of sewer system is found in older urban areas in the United States. Most of the time, a combined sewer system can transport all of its wastewater to a sewage treatment plant without incident. At the waste treatment plant, the wastewater receives appropriate treatment before being discharged (released) as effluent to a body of surface water. However, during periods of heavy rainfall or excessive snowmelt, the wastewater volume in a combined sewer system may exceed the capacity of that sewer system or its treatment plant. Anticipating such conditions, engineers designed the combined sewer systems to overflow occasionally and to discharge any excess wastewater directly to nearby streams, rivers, or other surface water bodies. Sanitary engineers refer to these overflows as *combined sewer overflows* (CSOs). They represent a major pollution concern in many American cities because CSOs contain not only storm water but also untreated human and industrial wastes, toxic materials, and debris.

Urban planners now prefer to use separate sanitary and storm water sewer systems to avoid the environmental consequences of CSOs. During periods of heavy rainfall, the separate storm water sewer carries the runoff to a nearby stream or water retention pond. The sanitary sewer system does not receive any storm water discharge. Instead, it remains dedicated exclusively to the transport of human and industrial wastes to the treatment plant, thereby avoiding precipitation-related raw sewage overflows.

It is very easy to overlook the importance and convenience of a safe municipal drinking water supply and a dependable sanitary sewer system. After all, the concepts have been around since ancient times, but the National Academy of Engineering (NAE) listed the modern water supply and distribution system as one of the greatest engineering achievements of the 20th century. A panel of distinguished engineers ranked this achievement fourth overall, placing it just behind electrification, the automobile, and the airplane and ahead of such other marvelous technical achievements as the computer, spacecraft, and the Internet.

## IRRIGATION

Scientists define irrigation as the human-controlled application of water in dry regions to assist in the growth of agricultural crops. Around the globe, people commit about 60 percent of all the world's freshwater withdrawals to irrigation. Large-scale farming could not grow food to support the world's ever-growing human population without the assistance of modern irrigation. In parts of California and Arizona, irrigation makes the desert bloom and a rich agricultural industry possible. According to agricultural scientists, it takes about 25 gallons (95 L) of water to grow an ear of corn; some 65 gallons (246 L) to produce a gallon (3.79 L) of milk;

and 100 gallons (379 L) to grown a watermelon. A two-pound (1-kg) loaf of bread requires about 1,000 gallons (3,785 L) of water.

Irrigation has been around since the start of civilization and has played a key role in establishing the stable food supplies that enabled human development beyond nomadic existence. Irrigation techniques in ancient civilizations were relatively simple, mostly involving pouring water on the fields. Ditches were dug to divert streams and simple rock and soil dams constructed to create modest-sized reservoirs. In approximately 250 B.C.E., Archimedes' screw appeared and began improving how humans moved water from nearby streams and lakes to thirsty fields throughout the Mediterranean basin. Because of its utility and simplicity, people still use the ancient device to move water, especially in certain underdeveloped regions of Africa and Asia.

Today, engineers have developed more efficient, mechanized ways of irrigating arid fields, but there is still room for improvement. When people use drinking water in their homes or businesses, about 90 percent is eventually returned to the immediate environment, where it replenishes local water sources. In contrast, about 50 percent of the water used for irrigation is effectively lost: by evaporation into the atmosphere, by the transpiration of plants, or by seepage into the ground during transport due to leaky pipes.

Today, there are many different approaches to more efficient irrigation. Each technique has distinct advantages and disadvantages. Taken as a whole, modern irrigation technologies allow a relatively few farmers to produce very large quantities of food on semiarid land in support of many other people. In sprinkler irrigation, for example, water is applied by means of perforated pipes or nozzles, which are operated under pressure in a manner that forms an appropriate spray pattern. Drip irrigation involves the use of special perforated pipes or porous tubing that apply water very gradually to the root zone of the plants being cultivated. Center-pivot irrigation systems often use a very long (about 1,250 feet [381 m]) traveling radial arm that is equipped with assemblies of piping and sprinklers. The giant mechanical arm slowly rotates around its pivot point, delivering moisture to the crop being cultivated. When viewed from above, center-pivot irrigation systems create a pattern of very distinctive crop circles.

To conserve water, farmers often capture and reuse any excess water that runs off cultivated fields as a result of irrigation practices. They capture this runoff in ponds and then use it during the next irrigation cycle.

Handline sprinkler irrigation providing precious drops of water to germinating crops in Yuma, Arizona *(USDA)*

## WATER TRANSPORT

The availability of water shaped and molded ancient civilizations. As people learned to manipulate water, they began to enrich their daily lives. Irrigation techniques assisted early farmers, and wells and simple water storage systems provided water for drinking, cooking, and personal hygiene. Early peoples also discovered how to use simple rafts and dugout canoes to travel across inland waterways.

Over time, the inhabitants of early civilizations learned how to construct ships that used both human muscles (oars) and wind power (sails) for propulsion. The Egyptians transported cargo along the Nile River using various types of barges and sailing ships. Then, as now, water transport offered an efficient way of moving large quantities of materials.

In about 1500 B.C.E., the Phoenicians emerged as the first great maritime trading civilization within the Mediterranean basin. From the coastal regions of what is now Lebanon, Phoenician sailors traveled across the Mediterranean in well-designed human-powered sailing vessels referred to by naval architects as *biremes.* The bireme had two sets of oars on each side of the ship and a large square sail. The ship's name results from a combination of *bi* (meaning "two") and *reme* (meaning "oars"). The Phoenicians were not only master shipbuilders and skilled traders, they also developed the early alphabet upon which present-day alphabets are based. The ancient Greeks and later the Romans improved the design of the Phoenician bireme. The trireme (three rows of oars on each side) emerged as the dominant warship of the Mediterranean basin. While the Dark Ages enveloped most of western Europe, Viking long ships began departing Scandinavian waters and prowling across the Atlantic Ocean. Looking for trade or plunder, Norse sailors ventured far up many great European rivers and made daring sorties into the Mediterranean Sea.

As the nations of western Europe began their ambitious programs of worldwide exploration in the 15th century, their military and commercial interests encouraged the design of improved sailing ships. A major breakthrough in water transport took place in 1807, when the American engineer and inventor Robert Fulton (1765–1815) inaugurated commercial steamboat service. On August 14, 1807, Fulton's steamboat called the *Clermont* made its journey up the Hudson River from New York City to Albany, demonstrating the great potential of steam-powered ships. A similar milestone in marine engineering occurred on January 17, 1955, when the world's first nuclear-powered ship, the submarine USS *Nautilus*

The merchant vessel *M/V Ever Ulysses* is a modern container ship. *(NOAA)*

(SSN-571), put to sea for the first time and its captain sent back the historic message "Underway on nuclear power."

Water transport represents a key component in the complex transportation system that supports the American economy. Included in the marine portion of this transportation system are a network of navigable inland waterways, ports, and cargo transfer hubs that link merchant ships to rail, air, road, and pipeline systems. According to statistical data from the U.S. Department of Commerce, the United States is the world's leading maritime trading nation. More than 78 percent of American overseas trade by volume and 38 percent by value comes and goes by ship. These data include 9 million barrels of imported oil daily. American businesses rely on water transport to access suppliers and markets worldwide. An average-size modern cargo ship can carry as much containerized cargo as 1,500 railroad cars or 6,000 semitrailer trucks.

Despite its great advantages, transportation by water has encountered one major geographical disadvantage throughout human history. Whether they were traveling by sea or along inland waters, early mariners were constrained by the natural routes and depths of the waterways they moved across. In the case of inland waterways involving rivers, the travelers often encountered natural obstacles such as waterfalls, rapids, and shallow waters. Hence, they would have to unload the cargo and take it, along with their boats, around the obstruction. Following the time-consuming and troublesome process of portaging, the travelers would continue their journey on the river. Similarly, two large bodies of water might be separated by a relatively small piece of land that served as a natural barrier to continuous travel. Once again, mariners were faced with the choice of unloading cargo and carrying it overland to the other body of water.

People in early civilizations appreciated the great advantages of water transport compared to hauling large quantities of goods across land, so canal building emerged as an early large-scale engineering achievement of these civilizations. By constructing canals, ancient peoples were able to overcome natural obstacles that hindered development and trade along important waterways. Historic records and archaeological ruins suggest that the early Egyptians, Greeks, Romans, and Chinese all constructed canals to facilitate water transport.

Canals continued to play important roles throughout history. Three canals of recent historic importance are mentioned here: the Erie Canal, the Suez Canal, and the Panama Canal. Each represented a human-engineered manipulation of water that greatly facilitated marine transport and world trade.

On November 4, 1825, New York State's governor, DeWitt Clinton (1769–1828), formally opened the Erie Canal in a ceremony in New York City by pouring water from Lake Erie into the Atlantic Ocean. The famous canal system was an engineering marvel of the 19th century. By providing safe and inexpensive transport from the port of New York to portions of the United States west of the Appalachian Mountains, the Erie Canal opened up the world to the agricultural and domestic products of the American Midwest. Later in the 19th century, as trains arrived on the transportation scene, canals began to take a backseat to the versatility and speed of steam-powered locomotives. The creation of the Saint Lawrence Seaway in 1959 along the Canadian-American border further diminished the value of the Erie Canal. Today, the 524-mile (840-km) waterway across upstate New York is designated the Erie Canalway National Historic Corridor. It remains the most important canal system ever constructed within the borders of the United States.

The Suez Canal is a human-constructed, sea-level waterway that runs north to south across the Isthmus of Suez in Egypt, separating the continent of Africa from Asia. The approximately 105-mile (169-km)-long waterway connects the Mediterranean Sea with the Red Sea and represents one of the world's more strategic waterways. Although most of the canal is limited to a single lane of traffic, the waterway contains four lengthy bypasses that allow ships passage in both directions.

The idea of linking the Red Sea with the Nile River and/or the Mediterranean Sea dates back about 40 centuries. According to hieroglyphic records, in approximately 1900 B.C.E. the Egyptian pharaoh Senausert III constructed an east-west canal that linked the Nile River with the Red Sea. Over the years, the accumulation of silt caused the ancient canal to be abandoned and then reopened several times. During Napoleon Bonaparte's military campaigns in Egypt in the early 19th century, French engineers found remnants of the ancient waterway, and their discovery renewed interest in a canal that would directly connect Europe with the Indian Ocean.

Eventually, the French diplomat and engineer Ferdinand de Lesseps (1805–94) organized the development of the modern Suez Canal. The objective of the project was to link the Mediterranean Sea with the Red Sea and the Indian Ocean beyond. Construction started in 1859, and the canal was opened to shipping in November 1869. In 1875, the British prime minister Benjamin Disraeli (1804–81) moved to secure effective British control of the canal by purchasing the Egyptian government's shares in the Suez Canal Company. Following several international conflicts in the 20th century, the Suez Canal Authority of the Egyptian government now owns and operates the important international waterway.

In 1880, building upon his success with the Suez Canal, Ferdinand de Lesseps organized a French company to construct a sea-level canal across the Isthmus of Panama. The company encountered unanticipated construction difficulties and oppressive cases of disease, primarily malaria. By 1889, the French abandoned the project. Early in the 20th century, the American president Theodore Roosevelt (1858–1919) pursued the political, economic, and technical pathways necessary to construct a canal across Panama, a newly independent country carved out of Colombia in 1903. Roosevelt appointed the U.S. Army officer and engineer George Washington Goethals (1858–1928) the chief engineer on the American canal-building effort across Panama. Through a treaty signed with the

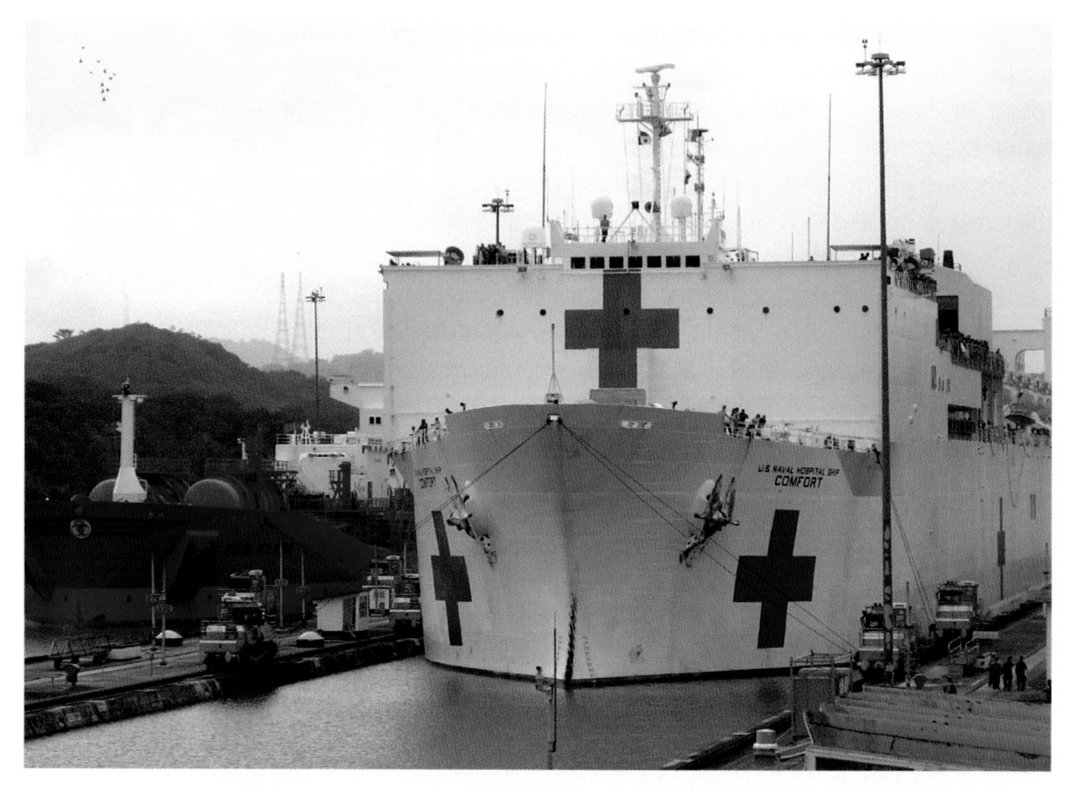


The hospital ship USNS *Comfort* (T-AH 20) makes its way through the Panama Canal as it travels to the Pacific Ocean. At the time, the *Comfort* was participating in Continuing Promise 2009—a four-month humanitarian and civic assistance mission involving seven countries throughout Latin America. *(USN)*

newly independent Panamanian government, the United States obtained a 10-mile (16-km)-wide piece of land across Panama, named it the Canal Zone, and governed the strategic strip of land as an American territory until 1977. More recent treaties returned the Canal Zone to Panama. Today, the Panama Canal Authority (PAC) operates and maintains the important human-made waterway.

Using ingenious construction techniques and bold designs, including high-lift locks, Goethals completed the Panama Canal in 1914, two years ahead of schedule. An engineering marvel, the waterway has a total length of about 50 miles (80 km) and consists of a series of human-engineered lakes, artificial channels, and three sets of locks. Almost a century after it opened, the Panama Canal remains a vital waterway that connects the Atlantic Ocean (Caribbean Sea) with the Pacific Ocean. The canal enjoys more ship traffic (some 15,000 vessels annually) than ever anticipated by its builders.

There are some lingering environmental and technical issues. For example, with the existing system of locks, each time a ship transits the canal, about 52 million gallons ($200 \times 10^6$ L) of freshwater, primarily from Lake Gatún, flow into the ocean. Engineers are now investigating the use of new locks with water-saving basins to preserve the freshwater resources of the canal's watershed. Also, the sizes of many modern naval vessels and cargo ships now exceed the canal's capacity. Expansion of the waterway and its system of locks appears appropriate if the canal is to maintain its strategic value as well as its market share of global shipping traffic.

## HYDROPOWER

Since ancient times, people have constructed dams for a variety of reasons. These include the impoundment of water in reservoirs for drinking and irrigation and the creation of artificial lakes behind the dams for recreation, the diversion or control of flood waters, and the harnessing of flowing water's energy. Engineers define hydropower as the use of flowing water to power machinery or to generate electricity (hydroelectric power). Contrary to popular belief, not all dams are used to produce electric power. Of the approximately 80,000 dams in the United States, only 2,400 generate electricity. The vast majority of American dams support other uses, such as creating reservoirs for drinking water supplies and irrigation, flood control, recreation, and the production of human-made agricultural ponds for livestock.

More than 2 millennia ago, the ancient Greeks used hydropower to turn waterwheels to grind wheat into flour. The Romans imitated these practices and constructed water-powered mills in appropriate locations throughout their empire. In the Greco-Roman era, waterwheels of various designs and complexities served two basic functions: to grind grain or to lift (pump) water in support of irrigation practices. Following the Dark Ages, waterwheels reappeared throughout medieval Europe, many associated with monasteries. Clever arrangements of wooden gears allowed hydropower to support sawing operations (sawmills) and other labor-intensive mechanical activities.

At the start of the First Industrial Revolution, many of the late 18th-century British textile mills were located near dependable sources of flowing water, but the arrival of commercially viable steam engines made the

## HOOVER DAM

Aerial view of Hoover Dam and a portion of its large reservoir, Lake Mead *(USDA)*

Spanning the Colorado River between Arizona and Nevada about 30 miles (48 km) southeast of Las Vegas is an engineering wonder of the modern world known as Hoover Dam. Named after the 31st American president, Herbert Hoover (1874–1964), this massive concrete arch-gravity dam impounds an enormous quantity of water using gravity and the forces generated by its horizontal arch design.

Hoover Dam is among the world's largest. The U.S. Bureau of Reclamation began construction of this magnificent structure in 1931, completed it in 1936, and continues to operate the facility. Workers used more than 3.25 million cubic yards ($2.46 \times 10^6$ $m^3$) of concrete to build the dam, which rises 726.4 feet (221.4 m) from the foundation rock to the roadway on its crest. The maximum hydrostatic (water) pressure at the base of the dam is 312.5 psi (2.155 MPa).

Lake Mead is the reservoir created by Hoover Dam. With a surface area of approximately 250 square miles (640 $km^2$), it is the largest reservoir in the United States and contains, when full, $28.5 \times 10^6$ acre-feet (35.2 $km^3$) of water. Water impounded by Hoover Dam serves municipal drinking water needs; supports irrigation-based agriculture in California, Arizona, and New Mexico; and generates electric power. The dam has an installed hydroelectric generating capacity of 2,080 megawatts (MW). On average, Hoover Dam generates about 4 billion kilowatt-hours of electricity each year for use in Nevada, Arizona, and California—enough to serve 1.3 million people. The dam supports flood control, helps regulate the Colorado River, and creates the Lake Mead National Recreation Area.

location of new factories independent of flowing water sources. Despite the technology transition, waterwheel-powered mills remained important in Great Britain, on the European continent, and in North America. For example, the first paper mill constructed on the Pacific Coast of the United States was located in Lagunitas, Marin County, California. Built in 1856, this paper mill originally used water power but later switched to steam engines.

Toward the end of the 19th century, the marriage of water power and electricity took place and changed the world. One historic example will highlight the intellectual excitement that accompanied the arrival of this important technical milestone. The visionary American engineer and entrepreneur George Westinghouse (1846–1914) helped create the modern electric power industry by financially supporting the Croatian-born, Serbo-American electrical engineer Nikola Tesla (1856–1943) and his development of alternating current (AC) generators, motors, and transformers. In 1895, Westinghouse's company won a coveted contract to use Niagara Falls to generate electricity and to deliver the hydropower-generated AC electricity to the city of Buffalo, New York, located some 22 miles (35 km) away.

There are three basic types of hydroelectric facilities: impoundment, diversion, and pumped storage. The most common type of hydroelectric power plant is an impoundment facility such as Hoover Dam. As shown in the accompanying illustration, an impoundment dam stores river water in a reservoir. As water is released from the reservoir through a penstock, it spins a turbine-generator system to produce electricity. The electricity is then transmitted away from the dam by high-voltage, long-distance power lines. Dam operators can release water according to electric power demands, or they can release water at a controlled rate so as to maintain a specific water level range in the reservoir. The turbine-generators in a typical hydroelectric dam are quite large.

In a diversion hydroelectric facility, engineers channel just a portion of the river's flow through a canal or penstock. Since a dam is typically not required, engineers sometimes refer to this type of hydroelectric plant as a run-of-river facility. Recognizing that water is much easier to store than electricity, engineers devised another type of hydroelectric plant, the pumped storage plant. During periods of light electric power demand, the facility uses some of its surplus generating capacity to pump water from a lower reservoir back into an upper reservoir. During periods of increased electric demand, the previously pumped water is released through penstocks to turbine-generator systems to produce electricity. This two-reservoir system represents a convenient way to store energy in the form of water at a higher gravitational potential until needed to satisfy peak electric loads.

Energy experts regard hydroelectric power generation as a nonpolluting, renewable source of energy. While such a facility does not release air pollution or put toxic materials into the water, there are still some potential adverse environmental effects that need to be assessed. The construction of large dams restricts the flow of natural, wild rivers and blocks the migration of fish such as salmon that must swim upstream to their spawning grounds. One partially successful solution is the use of properly designed fish ladders. Another environmental consequence of hydroelectric generation is that the reservoir created behind a dam floods the land,

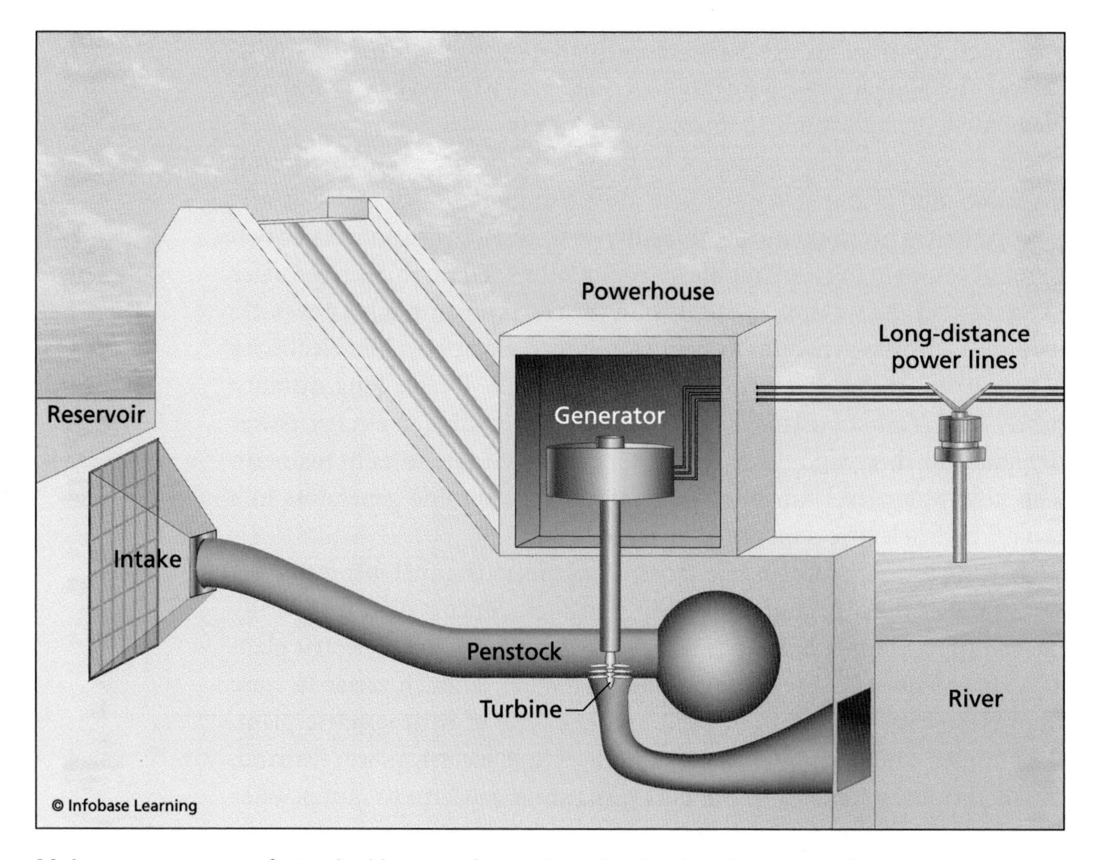


Major components of a typical impoundment-type hydroelectric power plant

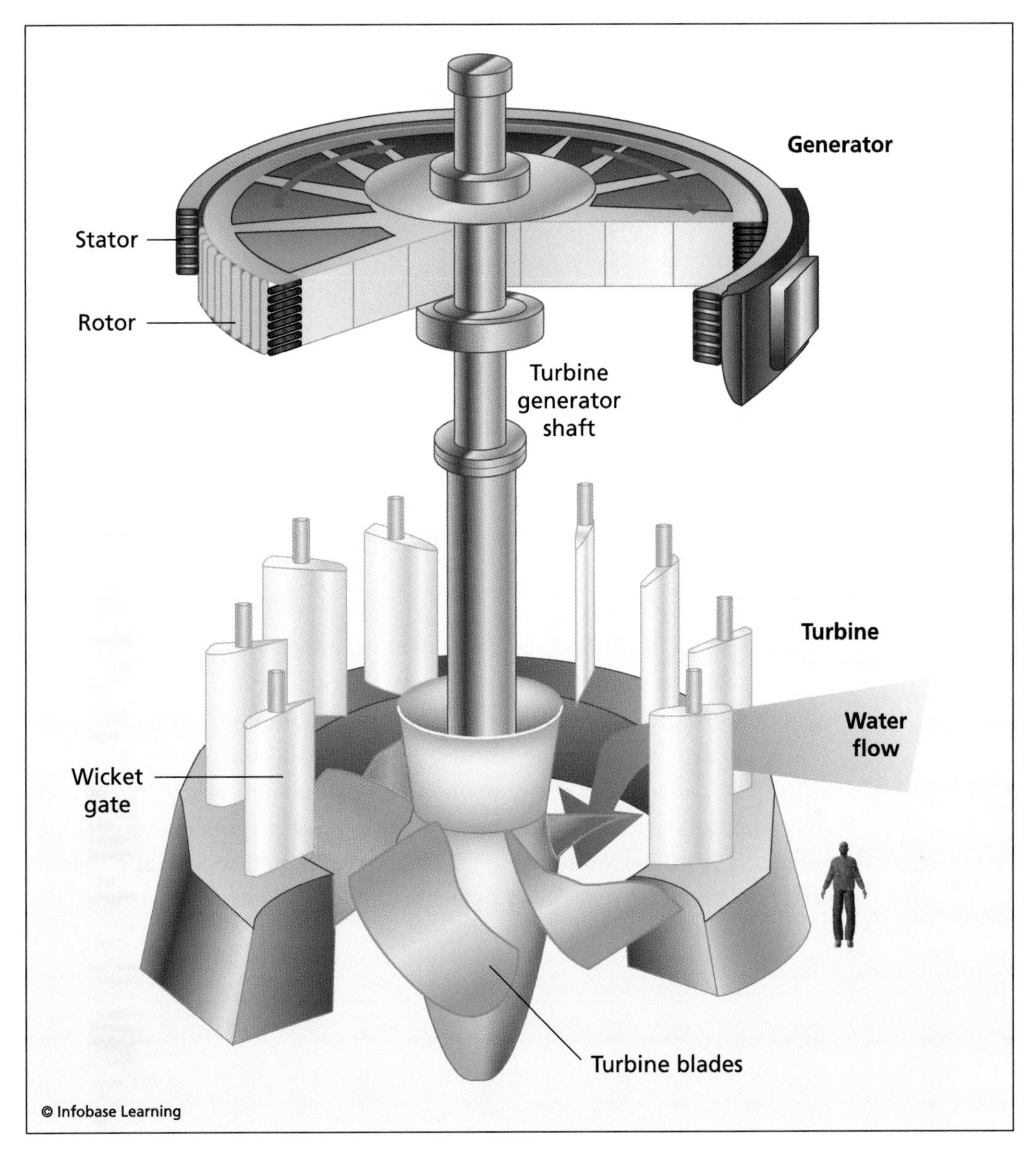


Components of a typical hydroelectric turbine-generator unit *(Modeled after U.S. Army Corps of Engineers)*

removing it from use for agriculture or for habitation by people or wildlife. Finally, a dam failure can release a flood of water downstream. Engineers and environmental planners carefully evaluate the major benefits and risks associated with each new hydroelectric project.

# CHAPTER 7

# Liquid Refreshments

**This chapter describes some of the more popular liquid refreshments** that people around the world enjoy daily. Food service industry personnel often use the term *beverage* to identify all liquid refreshments except water. Focus is given to the history, general properties, and characteristics of tea, coffee, beer, wine, alcohol, and carbonated beverages. Orange juice and milk are also mentioned. Because beverage preferences vary significantly from person to person and from society to society, the chapter strives to maintain objectivity by neither advocating nor discouraging the use of any of the beverages noted here.

## TEA

Next to water, tea is the most widely consumed liquid in the world. Starting with the Chinese writer and tea sage Lu Yu (733–804 C.E.) and his comprehensive treatise on tea (*Tea Classic,* or *Ch'a Ching,* published in about 780 C.E.), individuals have been writing about the botanical, pharmacological, and medical aspects of tea. For centuries, people have been boiling water to make tea. That simple aspect of tea preparation probably saved millions of lives during cholera epidemics, typhoid fever outbreaks, and other pandemics caused by waterborne pathogens.

Tea is a product of the plant *Camellia sinensis,* an evergreen shrub that grows in a variety of climates ranging from subtropical to tropical. Left undisturbed, the tea plant will grow into a small tree, but to make leaf plucking easier, the shrubs cultivated at tea plantations are usually pruned to waist height. There are three major kinds of tea, all prepared from the same plant: black tea, oolong tea, and green tea. All tea starts out as a green leaf of the *Camellia sinensis* plant. How the leaves are picked and processed establishes the type of tea that ultimately results.

Black tea leaves are fully fermented. Agricultural workers pick the leaves and then allow them to wither. After picking, the leaves of the tea bush rapidly wither and oxidize. As their chlorophyll breaks down, the leaves release tannins and become progressively darker. Without the release of the tannins, the tea would lack its characteristic full-bodied flavor and color. When a tea leaf is submerged in hot water, it brews an astringent (tart) flavor that is characteristic of tannins. (Wine, especially

## CAFFEINE

Caffeine is a mild natural stimulant found in tea, coffee, and cola soft drinks. While analyzing coffee in 1819, the German chemist Friedrich Ferdinand Runge (1795–1867) became the first scientist to isolate the caffeine molecule ($C_8H_{10}N_4O_2$). When purified, caffeine forms a very bitter, white, crystalline powder. Once ingested by human beings, caffeine increases the heart rate, stimulates the central nervous system, and has mood altering (psychotropic) properties.

Many Americans depend upon a cup of coffee each morning to obtain the mental alertness ("caffeine jolt") considered necessary to start their day. One cup of strong coffee or three to five cups of tea provide the average adult an effective dose, about 200 milligrams, of caffeine. Some people obtain a dose of caffeine by eating chocolate, especially dark chocolate, or by taking tablets prepared as antidrowsiness medications. Physicians generally consider caffeine safe for people when they ingest the chemical in moderate doses (on the order of 200 mg per day). However, continued consumption over long periods of time can lead to caffeine dependency. An excessive dose of caffeine in a short period may also cause caffeine intoxication. For an adult, the lethal ingested dose of caffeine ranges from 13 to 19 grams. Caffeine is very toxic to household pets, especially dogs. The chemical is also a natural pesticide, capable of paralyzing and killing certain insects as they feed on caffeine-bearing plants.

A glass of Turkish tea sits on a table at Incirlik Air Base, Turkey. Turkish tea is always served hot and in special tulip-shaped teacups so people can hold them easier and feel the warmth of the tea with their hands. *(U.S. Air Force)*

red wine, also contains tannins.) Next, workers roll the withered leaves to release their juices and then allow them to ferment. Finally, black tea leaves are oven-dried before packing. Black tea has a dark color and fuller taste. It also contains more caffeine than the less-oxidized oolong and green teas.

Oolong tea leaves are partially fermented. Workers apply the same process they use in preparing black tea, but they substantially shorten the fermentation time prior to packing. This semioxidized tea is commonly served in Chinese restaurants. The color and taste of oolong tea lies between that of black tea and green tea.

Green tea leaves do not receive any fermentation. Workers heat the leaves immediately after picking to prevent fermentation. The leaves are then rolled, dried, and packed. Green tea has a pale yellow color and less complex taste because of a relative absence of tannins.

Persons involved in the tea industry often use several additional terms to describe the size and age of black tea leaves. They refer to larger leaves as orange pekoe, pekoe, and pekoe souchong. They describe smaller, broken leaves as broken orange pekoe, broken pekoe, broken pekoe souchong, broken orange pekoe fannings, and fines (tea dust). Tea industry personnel use a similar collection of terms to describe the age of a tea leaf. From the youngest to the oldest, a leaf's age is described as flowery pekoe, orange pekoe, pekoe, souchong first, souchong second, Congou, and Bohea. By using different leaf sizes and ages, tea merchants create blends of teas. A popular commercial blend, orange pekoe tea, consists of young, small leaves.

There are many apocryphal legends about the origins of tea drinking in China several millennia ago. From China, the use of tea spread throughout Asia and then into the rest of the world. Sometime around 600 C.E., Buddhist monks introduced the Japanese to tea. At first, tea served as the beverage of the religious and noble classes, but later the general population in Japan began to enjoy the beverage. Not only was tea eventually cultivated in Japan, green tea became the social beverage of choice for most cultured Japanese citizens. The Japanese tea ceremony evolved during the practice of diplomacy in feudal Japan and remains an integral part of contemporary culture.

As early as the 13th century, European explorers and missionaries, such as the Venetian traveler Marco Polo (ca. 1254–1324), visited China and discovered tea. In the early part of the 17th century, a Dutch merchant ship brought green tea leaves from China back to Holland. Tea began to spread throughout Europe. British coffeehouses introduced tea as an alternate beverage in the 1650s. Due to intense commercial promotion, British East India Company tea soon became the favored beverage in Great Britain and many of its colonies, including those in North America.

European colonial expansion included the spread of tea cultivation, an activity filled with stealth and political maneuvering. To circumvent dependence on Chinese-produced tea, agents of the British East India Company managed to steal live tea plants and smuggle them out of China. This illicit botanical contraband gave rise to tea cultivation in India and Sri Lanka (formerly Ceylon). (As discussed shortly, similar acts of subterfuge established coffee cultivation around the world.) Tea became a very important trading commodity and assisted in creating the global dominance of Great Britain by the end of the 18th century. Conflict with the British over tea also helped inflame the spirit of independence among the American colonists.

Up until 1773, American colonialists were primarily tea drinkers. Following the Boston Tea Party at the end of that year, many Americans regarded tea drinking as disloyal to the spirit of independence that was then gripping the emerging young nation. The politically motivated transition from tea to coffee was eagerly supported by Dutch and French coffee merchants, who recognized the opportunity and began flooding the American market with inexpensive coffee beans. A combination of patriotism and economics encouraged Americans to become a coffee-loving nation.

According to data from the Food and Agricultural Organization (FAO) of the United Nations, world tea production continues to grow. Although black tea still dominates global tea production, FAO projections to the year 2017 indicate that world green tea production is expected to grow at a considerably faster rate than black tea—namely, 4.5 percent annually, compared to 1.9 percent for black tea. China is the top tea-producing country, followed by India, Sri Lanka (formerly Ceylon), Kenya, and Turkey.

## BOSTON TEA PARTY

On December 16, 1773, a band of American patriots, the Sons of Liberty, disguised themselves as Mohawk Indians and boarded three British merchant ships anchored in Boston Harbor. The *Dartmouth, Eleanor,* and *Beaver* were laden with crates of surplus tea that the financially struggling British East India Company wanted to dump on the American market at an artificially low (without duty or tariff) price. In protest that night, the Americans dumped hundreds of crates of tea into Boston Harbor without inflicting any damage to the ships or their crews. The event triggered a boycott of British tea throughout the colonies and encouraged the call to revolution against the British Crown.

The American colonists were already agitated by a series of extremely unpopular taxes and commercial tariffs that had already been imposed upon them by King George III. They regarded the Tea Act of 1773 as the final straw. The British strategy under this act was to undersell American merchants, who were already selling less expensive tea smuggled from Holland, and to gain a complete monopoly on the American tea trade. Reprisal by the British government for the Boston Tea Party was swift and harsh, including the closure of Boston Harbor.

## COFFEE

This section discusses coffee, another brewed beverage enjoyed by millions of people around the world each day. Americans consume more than 2.5 billion pounds ($1.14 \times 10^9$ kg) of coffee beans each year. Coffee beans are actually the seeds of the cherrylike berries of an evergreen tropical bush (genus *Coffea*) that is now cultivated in more than 80 tropical countries around the world. Although there are approximately 60 species of coffee plants, the two best known are the *Coffea arabica* and the *Coffea robusta.* Arabica trees yield better beans and account for about 70 percent of the world's annual coffee harvest; robusta trees are hardier, yield harsher beans, and account for about 30 percent of the annual coffee harvest.

The story of coffee is filled with apocryphal legends and tales of international intrigue. The original arabica coffee trees grew wild in ancient Abyssinia (modern Ethiopia). According an apocryphal tale, a goatherd named Kaldi observed the stimulating effect the berries of the coffee trees had on his goats. Word of his interesting discovery soon spread among the Abyssinians, but instead of using the seeds (beans) inside the cherrylike berries of the coffee shrub to brew a stimulating beverage, they chose to grind up the coffee beans and mix the powdery substance with butter or animal fat to create an energy-providing paste. About 850 C.E., Abyssinian farmers began cultivating arabica coffee trees. Arab traders soon discovered the stimulating properties of the coffee bean and carried some plants back to Yemen for cultivation.

In about 1000 C.E., Muslims in Yemen began brewing roasted coffee beans. Because the Islamic religion forbids the use of alcohol, brewed coffee, sometimes referred to as the wine of Arabia, became a religiously acceptable substitute stimulus. Arab merchants controlled the coffee trade by selling only roasted coffee beans. Trade in live coffee plants and botanically viable seeds was strictly forbidden. In 1536, the Ottoman Turks conquered Yemen and were instrumental in spreading coffee across their vast empire, including North Africa and into Europe. However, no coffee seed sprouted outside of Africa or Arabia until the bold escapades of several European merchants in the 17th century.

While trading with the Ottoman Turks in 1615, Venetian merchants encountered coffee and introduced the beverage to western Europe. The following year, Dutch merchants smuggled live coffee plants out of Arabia. As a result of this clandestine caper, the Dutch started coffee plantations in other parts of the world such as Ceylon (modern Sri Lanka) and the island of Java (Indonesia). In 1706, the Dutch transported a young coffee tree from Java and delivered it to the botanical gardens in Amsterdam, Netherlands. As the tree flourished, botanists carefully propagated it. Dutch merchants then shipped descendants of this tree to Dutch colonies in Suriname (South America) and the Caribbean. Dutch government officials also began bestowing live coffee plants on certain European aristocrats. In 1714, King Louis XIV (1638–1715) of France placed his coffee tree under guard in the Royal Botanical Garden (Jardin des Plantes).

Several years later, a French naval officer, Gabriel Mathieu de Clieu, wanted to imitate the Dutch by establishing coffee plantations on the island of Martinique in the eastern Caribbean. Denied official permission, he persisted in his vision. Employing romance and political favors within the French court, he eventually obtained an illicit plant clipping from the well-protected coffee tree in the Royal Botanical Garden. Despite a perilous journey to the New World in 1723, the sprout arrived alive in

Martinique. Cultivated under armed guard, this single tree produced a family of millions of coffee trees within half a century.

As interest in coffee grew in Europe, the governments of Holland, France, Spain, and Portugal became eager to further expand coffee cultivation in their far-flung tropical colonies. In 1727, the Portuguese colonial government in Brazil sent a military officer, Francisco de Melo Palheta, to French Guiana to mediate a border dispute. His real mission was to smuggle live coffee plants from heavily guarded French coffee plantations. He so charmed the French governor's wife that she gave him some coffee plant seeds hidden in a farewell bouquet. From these seeds arose the millions of coffee trees that made colonial Brazil (independent in 1822) the world's leading coffee producer. Spanish and Portuguese coffee plantations appeared throughout the tropical regions of Central and South America.

Unlike the other European powers, Great Britain did not actively participate in the fierce coffee cultivation race of the 17th century. The British

## ORANGE JUICE

Orange juice is a classic American beverage, found in approximately 70 percent of the households in the United States. Some 80 percent of the orange juice sold in the United States is made from oranges grown in Florida, second in the world, behind Brazil, in orange juice production. In the 16th century, Spanish explorers planted the first orange trees around St. Augustine, Florida.

The American citrus industry has developed techniques to make a supply of high-quality orange juice available throughout the year. Oranges are picked when ripe, washed, and then squeezed to extract the juices. Seeds and particles are strained out. Citrus industry workers then pasteurize orange juice to ensure its safety. During pasteurization, the juice is heated to an elevated temperature for a short period of time. Orange juice containers labeled "not from concentrate" indicate that the fruit has been squeezed; the juice pasteurized, chilled, and blended for consistency and taste; and the result then packaged in cartons or safe-stored for later packaging. Orange juice containers labeled "from concentrate" indicate that the juice has been squeezed and the water evaporated, thereby producing a concentrated form similar to the one found in cans in the frozen food section of a food store. Upon the addition of water at processing plants, the concentrate becomes juice again, undergoes pasteurization, and is packaged.

people did, however, enjoy coffee, and by the mid-17th century, London had more than 300 coffeehouses, which facilitated social networking. Later that century, shrewd advertising campaigns by the British East India Company slowly weaned the British population away from coffee and made tea the national beverage. In the late 1800s, British explorers discovered the *Coffea robusta* tree growing wild in the Congo (Central Africa), so British coffee plantations began cultivating this type of coffee tree in Kenya and Uganda. Today, robusta trees are also cultivated in Indonesia and Southeast Asia. In the 19th century, French missionaries introduced coffee cultivation in Vietnam, while German immigrants started sophisticated coffee cultivation projects in Guatemala.

When the cherrylike fruit of the coffee tree is ripe, workers harvest it by hand (select picking) or mechanically (strip-picking). The coffee seeds (beans) are removed and processed. The coffee bean contains two main chemical compounds that account for its characteristics as a brewed

By law in the United States, the term *100 percent orange juice* indicates that the juice has been prepared only from oranges. Natural orange components such as pulp may be part of the liquid sold as pure juice. The term *USDA Grade A* indicates that the orange juice meets the premium standard of quality established by the U.S. Department of Agriculture for color, flavor, and absence of defects.

Orange juice is a healthy, naturally sweet beverage that contains a variety of vitamins and minerals. For example, vitamin C is an important antioxidant and helps support a healthy immune system. An eight-fluid-ounce (240 mL) glass of orange juice contains 82 mg of vitamin C—at least 100 percent of the recommended daily value.

**Fresh squeezed Florida orange juice** ***(USDA)***

beverage. Caffeine provides a stimulating effect, while caffeol, a volatile oil, supplies flavor and aroma. Roasting unlocks the characteristic flavor of the coffee bean. Seven minutes of roasting produces a light coffee roast; nine to 11 minutes creates a medium (or breakfast) roast; 12 to 13 minutes of roasting creates a dark (Viennese or French) roast; and after 14 minutes, the beans get very dark and oily, yielding an espresso (Italian) roast.

Today, coffee beans are one of the world's most traded substances. The major coffee-producing nations in the world are Brazil (one-third of the world's coffee), Colombia, Indonesia (the largest producer of robusta beans), Vietnam, Mexico, Ethiopia (the natural home of the arabica tree), India, Guatemala, Côte d'Ivoire (the second-largest robusta bean producer), and Uganda. For some countries, coffee cultivation is a major part of the national economy. More than 15 million people in Ethiopia (25 percent of the population) derive their livelihood from the coffee sector. Coffee, especially robusta beans, accounts for 75 percent of Uganda's export revenue.

## BEER

Beer is the world's third most popular drink, after water and tea. It is also the oldest and most widely consumed alcoholic beverage. In its simplest modern definition, beer is any fermented, carbonated alcoholic beverage that has been brewed from malted barley (or other cereal) and flavored with hops. Agricultural scientists define malted barley as barley that has been allowed to sprout (germinate) to a certain degree and then dried. Barley is the seed of the barley plant, a grain similar in appearance to wheat. The process of germination converts the seed's starch into simpler fermentable sugars. Brewers add yeast to promote the fermentation process.

The practice of making beer is as old as (if not older than) human civilization. According to archaeological records, the Sumerians of ancient Mesopotamia were the first beer makers. Sometime around 7000 B.C.E., early farming practices in Mesopotamia yielded a variety of wheat that proved useful for both bread making and beer making. Although details of the actual discovery process remain unclear, apparently a batch of Sumerian bread may have gotten soaked with water and remained wet for several days. The arrival of airborne yeasts turned the soggy dough into alcohol. A brave Sumerian sampled the resultant mixture, experienced a pleasant sensation, and started the art of brewing. The Sumerians liked the beverage so much they created a beer goddess, Ninkasi. Beer allowed the Sumerians to store agricultural surpluses in beverage form to provide sustenance during lean growing times.

Beer brewed from malted barley *(USDA)*

As the Babylonians gained control of Mesopotamia and created their own empire, they also learned to how brew beer and produced several new types. The legend of the Babylonian hero Gilgamesh (ca. third millennium B.C.E.) describes the transforming powers of beer. The great Babylonian king and law giver Hammurabi (ca. 1792–50 B.C.E.) issued proclamations governing the daily rations of beer for his subjects. He based the beer rations on a person's social status. Then as now, civil servants received larger shares than ordinary workers.

Beer making also became an integral part of other early civilizations. The ancient Egyptians brewed beer in the Nile River Valley, using unbaked bread. Through trade, they passed brewing technologies and recipes to the ancient Greeks, who in turn taught the early Romans. Sometime during the Roman Republic (ca. 500 B.C.E.–30 B.C.E.), the Romans replaced beer with wine as their preferred alcoholic beverage. Drinking beer became viewed as an act of barbarians, such as the Teutonic tribes along the frontiers in Gaul. At the time, Teutonic beer was very cloudy, lacked foam, and did not store for long. The imperial Romans preferred to spread vineyards throughout their portions of western Europe.

After the fall of the Roman Empire (ca. 500 C.E.), Christian monks began brewing beer in their monasteries. They learned how to produce a nutritious, pleasant-tasting alcoholic beverage that they could drink with their rather meager meals. Sometime in the early ninth century, German monks came upon the very important idea of adding hops to their beer. Since drinking beverages of any kind did not violate the strict fasting and dietary rules of most monastic orders, the monks could consume as much of the home brew as they wished. Religious historians suggest that many monks consumed up to 1.3 gallons (5 L) per day. (As discussed shortly, Christian monks also maintained vineyards and produced wines.)

Depending on local feudal laws, monasteries often ran monastic pubs where commoners and nobility alike could enjoy freshly brewed beer. Eventually, some of the dukes, princes, and kings wanted to extract more tax money from the monks for the privilege of selling beer to the local population. Feudal greed caused the closure of many monastic pubs. When beer produced by monks was no longer available in a locale, the peasants resorted to home brewing. Sometimes the results were acceptable, and other times the amateur beer maker (typically a housewife) unintentionally tossed deadly herbs into the brewing kettle for flavor. People in medieval Europe were very superstitious. Often, government officials would attribute failed (possibly lethal) brewing activities to the work of "beer witches" or "brew witches." Many beer witches were burned at the stake for their brewing mistakes.

One of the great milestones in beer making was the passage of the German Beer Purity Law (Reinheitsgebot) in 1516. Issued by Wilhelm IV of Bavaria (1493–1550), this law governed beer production in Bavaria for centuries until it was abolished as a binding obligation in 1986 by regulations of the European Union. It specified that beer could have only water, malted barley, and hops. Yeast was later added to the list of acceptable ingredients. This law shaped German beer making, which subsequently shaped modern beer brewing around the world.

During the Industrial Revolution, beer making transitioned from a cottage industry to a large-scale industrial activity. Advances in technology greatly assisted 19th-century brewmasters as they strove to produce consistently good quality beer for the working masses as well as the upper classes. Steam-powered breweries sprang up throughout Europe. At one of these in the United Kingdom, the British scientist and brewer James Prescott Joule (1818–89) conducted pioneering scientific experiments that changed the world of physics. Specifically, around 1845, Joule performed a precisely instrumented experiment that demonstrated the mechanical equivalence of heat. Scientists now call the SI unit of energy the joule (J) in his honor.

Another great scientist of the era, Louis Pasteur (1822–95), described the scientific basis for fermentation, wine making, and the brewing of beer. While carefully examining a contamination problem in alcoholic fermentation, Pasteur concluded that fermentation was a biological process carried out by microorganisms. He followed this important postulation, which scientists now call *germ theory,* with many elegant experiments that clearly demonstrated the existence of microorganisms and their effects on fermentation. His pioneering work even led to a patent from the U.S. government entitled "Improvement in Brewing Beer and Ale Pasteurization."

By 1870, large commercial breweries in the United States started using refrigerated railroad cars to keep their beer cold during shipping. Today, most of the world's beer is produced by several dominant multinational corporations, although brewpubs and small regional breweries also flourish. The basic techniques of modern beer brewing cross national and cultural boundaries.

Beer industry personnel divide beer into two main categories: pale lagers and regionally distinct ales. Modern-day beers have a typical alcohol content that ranges from 3 percent to 6 percent alcohol by volume, although some strong beers have much higher alcohol contents.

The mashing room at the Orion Brewery, Nago, Okinawa *(USMC)*

## MILK

Milk is the whitish liquid that is a produced in the mammary glands of all mature female mammals. After giving birth, the female mammal suckles (nourishes) its young by providing milk. Like other mammals, human mothers can nurture their infant children in this way. At some point in a child's development, cow's milk, goat's milk, or similar forms of animal milk become a primary source of nutrition. Fluid scientists describe homogenized milk as an emulsified colloid consisting of a water-based liquid in which globules of liquefied butterfat are uniformly dispersed.

In the United States, dairy farming is an important part of the agricultural industry. Most commercially available milk comes from breeds of cows genetically selected for milk production. Black and white Holstein cows make up more than 90 percent of the American dairy herd. Jersey cows are the second most popular, making up about 7 percent of the dairy herd. The milk from Jersey cows is especially suitable for the manufacture of cheese because it contains a higher percentage of fat and protein.

A visit to any modern food store reveals the wide variety of products that result from milk. These products include fluid (liquid) milk, cheese, butter, yogurt, and ice cream. There are several varieties of fluid milk: skim milk (0 percent butterfat), 1 percent, 2 percent, and whole milk (about 3.5 percent butterfat). Milk from a dairy cow typically averages 3.5 percent butterfat, so dairy farmers separate raw milk into skim milk and cream and then reblend to achieve the standard fat content for each dairy product. They use the extra cream to make other liquid milk products such as half-and-half or whipping cream or to manufacture butter and ice cream. In the United States, fluid milk is pasteurized by rapidly heating it to 162° to 167°F (72° to 75°C) for from 15 to 20 seconds and then quickly cooling it. This process kills potentially harmful bacteria. The American dairy industry also homogenizes fluid milk and fortifies it with vitamins A and D. In the homogenization process, fat droplets in the milk are uniformly dispersed so that they do not float to the top.

According to the U.S. Department of Agriculture, the average American consumes 25 gallons (95 L) of milk each year. Those 25 gallons (95 L) of milk can make 9 gallons (34 L) of ice cream, 25 pounds (11.4 kg) of cheese, or 11 pounds (5 kg) of butter. Most dairy farmers milk their cows twice each day. The average dairy cow yields a total of 6.5 gallons (24.6 L) of milk each day.

Lactose is the natural sugar found in milk and milk products. Some people have difficulty digesting lactose. Physicians refer to this condition as lactose intolerance.

Pale lagers are the most commonly consumed beer in the world and are generally based on the pale lager brewed in 1842 in the town of Pilsen (now located in the Czech Republic). In 1839, officials of Pilsen recruited the Bavarian brewer Josef Groll (1813–87) to work at the city-owned brewery. On October 5, 1842, Groll introduced the first batch of clear, golden pilsener beer. His product became an instant sensation and now represents the model of beer for millions of consumers.

Today, beer is commercially manufactured in several stages using malted barley, water, hops, and yeast. During the brewing process, the starch in the cereal, usually barley, is transformed into wort, a sugary liquid. Then, yeast stimulates the fermentation process that transforms wort into beer. Basically, microorganisms metabolize the sugars extracted from the malted barley and produce alcohol and carbon dioxide. Mashing is the first step, during which a brewer grinds malted barley and combines it with hot water. The resulting sweetened liquid (wort) is drained off and placed into a large kettle. Hops are added, and the wort is then boiled to provide flavor and aroma to the liquid. The amount of boiling and the quantity of hops added during the boiling process determine the overall taste of the resultant beer. After boiling, the brewer cools the liquid and adds yeast to promote fermentation. During fermentation, the hop-flavored wort becomes one of many types of alcoholic beverage known as beer.

Some brewers substitute wheat, rice, or other raw cereal as the starch source during the initial stage of the beer-making process. Generally, brewers use slow-acting yeasts to create clear (pale), dry beers called lagers, and fast-acting yeasts to produce sweeter, darker beers called ales. Commercial brewers often use pasteurization to prevent further fermentation of their beer (packaged in bottles, cans, or kegs) during storage after manufacture.

## WINE

This section describes the history of wine and its role in human society from the dawn of civilization through modern times. Wine is an alcoholic beverage made by fermenting the juice of grapes. One of the most basic ways to classify wines is by their color: red, pink (rosé), or white. Wines may also be nonsparkling or sparkling (carbonated). Oenology is the study of wine, from the ancient Greek word *oinos* (οινος) for wine. Viticulture means the cultivation of grapes, from Latin *vitis* for vine.

When and who made the first wine is lost in history. What fragmented evidence is available suggests that sometime between 6000 and 5000 B.C.E., Neolithic peoples in the eastern Mediterranean (possibly in

Caucuses or Iran) may have stumbled upon well-ripened (possibly rotting) wild grapes, consumed them, and discovered the alcoholic beverage being fermented within the juices of the fruit. Archaeological sites including evidence of crushed grapes also suggest that the earliest wine production in Europe probably occurred in Macedonia about 4500 B.C.E.

Once grape cultivation took place around the Mediterranean, it was a relatively easy step for early Bronze Age farmers (vintners) to connect ripened grapes and wine. Grapes have a natural chemical balance and can ferment completely without need of additional sugars, acids, enzymes, or other nutrients. Traders brought the cultivation of grapes from the Levant (modern Lebanon, Syria, Jordan, and Israel) to ancient Egypt around 3000 B.C.E. Hieroglyphics in Old Kingdom–period tombs (ca. 2600 B.C.E.) depict vineyards in the Nile River delta. Wine is also mentioned in the Bible (both Old and New Testaments).

However, it was the ancient Greeks who truly embraced wine making and spread the use of wine through their city-states and colonies across the Mediterranean. The epic poems *Illiad* and *Odyssey,* by the Greek poet Homer (ca. ninth century B.C.E.), clearly describe the use of wine by the legendary heroes of the Trojan War. The ancient Greeks passed wine making on to the Romans, who made wine an integral part of their culture. As the Roman Empire expanded, the Romans established vineyards in all suitable wine-producing regions of western Europe. They also improved wine-making techniques and cultivated new varieties of grapes.

A sample of red wine being poured for a guest at a winery in Northern California *(USAF)*

The Romans initiated the practice of storing wine in barrels rather than in animal skins or clay jars, called amphorae. By the first century C.E., the Romans were shipping barrels of wine from Italy to their provinces in Gaul (France), Britannia (England), Germany, and Spain. Romans in these provinces established their own vineyards, and their wines were soon competing favorably with those produced in Italy. France's dominant position as a major wine-making nation traces its origins back to Roman viticulture.

The Romans also introduced the use of glass bottles for special wines. Archaeologists discovered a glass wine bottle that dates back to about 325 C.E. at a Roman gravesite in a vineyard near Speyer, Germany. As indicated by this well-preserved ancient wine bottle, instead of corking their wine bottles, the Romans chose to float a layer of olive oil on top of the wine in order to keep it from oxidizing.

When the Roman Empire collapsed around ca. 500 C.E., the only stable political structure remaining in western Europe was the Roman Catholic Church. To ensure a supply of wine for the celebration of the Catholic Mass, monasteries throughout western Europe turned to viticulture and wine making. The vineyards in Burgundy, Champagne, and the Rhine Valley trace their heritage back to monastic wineries established during the Dark Ages. The Benedictine Order became the largest wine producer in France and Germany. Centuries later, the Benedictine monk Dom Pérignon (1638–1715) made important contributions to the production of wine in the Champagne region of France. Popular misconceptions about this influential monk are that he invented sparkling wine and/or champagne. His primary efforts were to prevent secondary fermentation episodes, which would break wine bottles in storage. He also introduced the important activity of blending grapes prior to pressing. In his honor, the Möet & Chandon firm named their famous champagne Dom Pérignon.

When Spanish explorers and missionaries arrived in the New World during the 16th century, they brought along the cultivation of wine. Two centuries later, the Franciscan friar Junipero Serra (1713–84) dotted the California landscape from San Diego to San Francisco with missions. In 1769, Friar Serra planted the first "mission grape" vineyard (a variety of *Vitis vinifera* originally from Spain) at the Mission San Diego de Alcalá. Vineyards at Serra's California missions provided a stable supply of sacramental wine for use in the Catholic Mass. The Franciscan missionaries also made a variety of wines (red and white, sweet and dry, and brandy) for table consumption. From these humble beginnings and the industrious efforts of other European immigrants in the 18th and 19th centuries, the world-famous California wine-growing industry emerged. The top 10 wine-producing countries in the world, in descending order, are France, Spain, Italy, the United States, Argentina, China, Australia, South Africa, Germany, and Portugal.

There are numerous types of wine, and it ranges in price from inexpensive to extremely expensive. Rare, superpremium wines are the most expensive of all foodstuffs. Wine making is both an art and a science. Winery personnel divide wine production into two basic categories: without carbonation, a process called still wine production, and with carbonation, the process that leads to sparkling wines.

After harvest and blending, workers crush the grapes and allow them to ferment. To produce a red wine, workers generally allow the pulp of red or black grapes to ferment along with the grape skins. They make rosé wine by controlling the amount of contact between the grape juice and the skins. Finally, they produce white wine by restricting contact between the squeezed (pressed) juices and the grape skins.

After a period of primary fermentation, winery workers transfer the maturing liquid to various vessels for a secondary fermentation process. Some red wines experience secondary fermentation in wooden barrels, while white wines are often placed in stainless steel vats during secondary fermentation. The length of time committed to secondary fermentation establishes the taste, body, and character of a particular wine. The secondary fermentation period can extend from months to years depending on the grapes being fermented and the characteristics of the wine maker's target wine. Upon the completion of secondary fermentation, the wine is ready for sampling, additional processing, and bottling. As part of the production of sparkling wines, wine makers add a little sugar before bottling so additional fermentation takes place inside the bottle, releasing the characteristic carbon dioxide bubbles that become entrained within the liquid.

## ALCOHOL

This section briefly discusses alcohol, specifically drinking alcohol. As mentioned in the previous sections, drinking alcohol, or ethanol ($C_2H_6O$), appears when malted barley or grapes ferment. In the former case, the resulting alcoholic beverage is beer, and in the latter case, wine. Since most yeasts do not function well when the alcohol content of a liquid rises above about 15 percent alcohol-by-volume (abv), wines and beers typically contain relatively modest quantities of alcohol. Most commercial beers produced in the United States have between 4 percent and 6 percent abv, while most wines, except fortified wines, range from 9 percent to 16 percent abv.

Chemists define distillation as the process in which a liquid is boiled and the resulting vapor condensed and collected. They refer to the collected vapor as the *distillate*. Both research chemists and chemical engineers use distillation to purify liquids and to separate mixtures of liquids. The refining of crude oil (discussed in the next chapter) is based on fractional distillation. Alcoholic beverages developed by distilling fermented beverages are sometimes referred to as *distilled beverages* or *spirits*.

The Arab alchemist Abu Musa Jabir Ibn Hayyan (ca. 721–ca. 815), also known by his Latinized name, Gerber, developed distillation equipment and then observed that distilled wine gave off a flammable vapor. This line of research was continued a few years later by another Arab alchemist, Muhammad ibn Zakariya Razi (825–925), who generally receives credit for the isolation of alcohol (ethanol) by distillation. In 1808, the Swiss chemist Nicolas-Théodore de Saussure (1767–1845) determined the chemical composition of ethanol.

Ethanol is a volatile, colorless liquid with a boiling point of 173°F (78.4°C) at atmospheric pressure. It burns with a smokeless characteristic blue flame. The complete combustion of ethanol results in the formation of water and carbon dioxide. Ethanol has applications in chemistry, industry, transportation (as an automotive or rocket fuel), medicine, and alcoholic beverages.

The amount of alcohol in a distilled beverage is commonly stated in two basic ways: alcohol-by-volume (abv) and proof. Distilled beverage merchants in the United States define *proof* as twice the percentage of alcohol-by-volume (abv) that a liquid contains at 60°F (15.6°C). A distilled beverage that is 100 proof has 50 percent abv. Conventional distilling techniques cannot produce an alcoholic beverage that is more than 191.2 proof (95.6 percent abv). Some commercially prepared distilled beverages are flavored, while others are not.

In the United States, a 12-fluid-ounce (350 mL)-serving of a typical commercially brewed beer contains about 0.6 fluid ounces (18 mL) of ethanol. This is comparable to the alcohol content of a five-fluid-ounce (150 mL) glass of wine or a 1.5-fluid-ounce (4mL)-serving of an 80 proof (40 percent abv) distilled spirit.

Governments around the world generally regulate the use of alcoholic beverages (beer, wine, and distilled spirits) within their political boundaries. If not forbidden for all by religious or political customs, the consumption of alcoholic beverages by people below a minimum age is usually restricted. Also, most responsible governments have enacted strict laws concerning impaired (drunk) driving. While alcoholic beverages may serve as an acceptable social lubricant and provide a certain level of personal relaxation and enjoyment, the excessive consumption of alcohol (a disease called *alcoholism*) can and has caused great harm within societies—past and present.

## CARBONATED SOFT DRINKS

Beverage industry workers define a soft drink as a beverage that does not contain alcohol. When carbon dioxide is dissolved in water, a mild acid called carbonic acid ($H_2CO_3$) forms. Sometimes called carbonated water, soda water, or sparkling water, this effervescent liquid produces a pleasant tingling sensation when consumed. The amount of carbon dioxide dissolved in the water determines the ultimate fizzy taste of the drink. Carbonated water forms the basis of carbonated soft drinks.

Some mineral waters are naturally effervescent. In 1767, the British chemist Joseph Priestley (1733–1804) discovered an artificial way of carbonating water. He suspended a bowl filled with water over a vat of beer at a local brewery in the city of Leeds, Great Britain. To his surprise, the water acquired carbonation and became a pleasant drink. At the time, Priestley referred to the then unknown gas responsible for the carbonation as "fixed air." Several years later, he placed some sulfuric acid on chalk to release carbon dioxide gas. He then dropped the bubbling chalk in a bowl of water and observed how it agitated the water as it dissolved. He described his pioneering carbonation research in a 1772 pamphlet entitled "Directions for Impregnating Water with Fixed Air."

During the STS 51-F mission of the space shuttle *Challenger* in 1985, astronauts experimented with carbonated beverages (specifically Coca-Cola and Pepsi Cola) in the microgravity conditions of orbital flight. This picture shows a fizzy droplet of Coca-Cola floating in microgravity. Since the bubbles of carbon dioxide are not buoyant, they remain randomly distributed in the drop of fluid, making it a somewhat foamy mess. *(NASA)*

Although Priestley never profited from his research with carbonated water, the German-born Swiss watchmaker and amateur scientist Johann Jacob Schweppe (1740–1821) did. He developed a process to manufacture carbonated water on an industrial scale based on Priestley's research. He founded the Schweppes Company in 1783 in Geneva, Switzerland, and then moved his business activities to London in 1792. The company's tonic water (carbonated beverage flavored with quinine) is the world's oldest commercially prepared carbonated beverage. Other popular Schweppes Company carbonated beverages included ginger ale and bitter lemon.

Entrepreneurs in the United States did not ignore the growing market for carbonated soft drinks. Responding to a prohibition on the sale of alcoholic beverages by the city of Atlanta in 1886, the American pharmacist John Stith Pemberton (1831–88) reformulated a version of his popular elixir, originally called French Wine Coca, substituted sugar syrup for wine, and gave his new carbonated beverage the iconic name Coca-Cola®. Since then, Pemberton's carbonated beverage and its commercial competitors have provided cool refreshment for millions of thirsty people around the world. In the 21st century, nutritional experts are recommending that schools reduce the availability of high sugar–content soft drinks as a way of combating the epidemic of childhood obesity now plaguing American youths.

CHAPTER 8

# Lethal Liquids

**Although many liquids are toxic to human beings and/or harmful to the** environment, this chapter focuses on two, petroleum and mercury, which are ubiquitous and essential. Petroleum is the lifeblood of modern civilization, while mercury has widespread uses in scientific instruments, personal electronics, street lights, fluorescent lamps, dry cell batteries, and even dental fillings. When properly used, controlled, and disposed of, both petroleum and mercury are marvelous technical servants of the human race, but when haphazardly released into the environment by accident or intention, they can cause serious harm. Volumes have been written about the environmental consequences of each of these liquids. This chapter can provide only a brief summary of what the liquids are, where they come from, how they are used, and what some of their potential environmental impacts are.

## ROCK OIL

The word *petroleum* is a combination of the ancient Greek words for rock and oil. Known to many ancient peoples because of seepage, crude oil is a naturally occurring liquid found in various underground geologic formations. Typically black or dark brown, crude oil consists of a complex collection of hydrocarbons, ranging from $C_5H_{12}$ to $C_{18}H_{38}$ in molecular composition. Engineers regard petroleum as an important

Working for the U.S. government, civilian contractors attempt to extinguish an oil fire in Kuwait during Operation Desert Storm (1991). In a defiant act of environmental terrorism, the former dictator of Iraq, Saddam Hussein (1937–2006), ordered his retreating army to set fire to Kuwait's oil fields. *(U.S.* Army)

primary energy source and term it a *fossil fuel* because of its prehistoric origins.

It is very common in daily language to interchange the words petroleum and oil. Almost everyone does it. Material scientists define *oil* as a substance that at temperatures near room temperature (68°F [20°C]) exhibits a viscous liquid state. Typically, an oily liquid does not mix with water. Scientists use the words *immiscible* (nonmixing) and *hydrophobic* to describe these characteristics. However, one type of oil is usually tolerant of other types of oil and will usually mix well with them. Scientists often use the term *lipophilic* (which literally means fat-loving) to describe this tendency. Chemists treat a liquid as lipophilic if it is able to dissolve much more easily in other oily organic compounds than in water. The word *oil* is actually a rather nonscientific term. People encounter all types of "oil" in daily life. There are cooking oils, such as peanut oil; animal fat oils, including grease; salad oils; fuel oils; body lubricants, such as suntan oil; and even fragrant holy oils (chrisms). In this chapter, the word *oil* means petroleum unless specified otherwise.

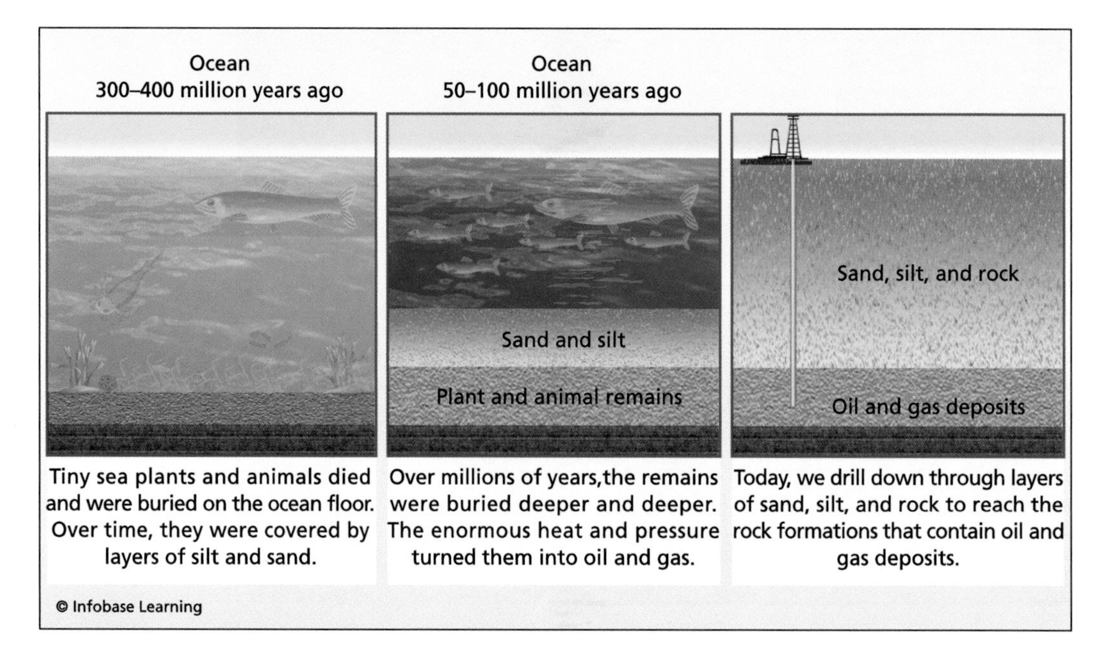


The origin of petroleum and natural gas *(Modeled after DOE)*

The illustration describes how petroleum and natural gas formed from the remains of ancient marine plants and animals that died millions of years before the dinosaurs roamed Earth. When these tiny prehistoric marine creatures and plants died, their organic remains settled on the ocean floor. (Scientists now speculate that coal formed mostly from the decay of ancient plants, while petroleum formed mostly from the decay of ancient animals, especially from their fats.) Over time, sand and silt covered the decaying material. As millions of years passed, the remains sank deeper into Earth's crust. Deep within the crust, the intense heat and pressure slowly converted the ancient organic remains into crude oil and natural gas. Methane ($CH_4$) and ethane ($C_2H_6$) are examples of natural gas. Crude oil is a smelly, light brown-to-black viscous liquid that has typically accumulated in underground pockets called reservoirs.

Geologists explore for oil reservoirs by studying rock samples taken from suspected oil-bearing regions. If the underground site appears promising, drilling operations begin. On land or at sea, engineers employ a structure called a derrick to house the pipes and tools going into the well. When finished, a successful, properly drilled well brings a steady flow of crude oil to the surface.

Prior to the advent of modern drilling technologies and pressure control equipment in the early decades of the 20th century, the rapid release of crude oil from an underground reservoir would often accompany a successful drilling operation. Oil industry workers coined the iconic expression *gusher* to describe the wild situation, when it quite literally "rained oil" all around the drilling site. The precipitation of crude oil continued until workers could successfully cap the new well.

Even today, a blowout can take place when the pressure-control equipment at an oil well fails. On April 20, 2010, the largest underwater blowout in history took place in the Gulf of Mexico off the coast of Louisiana. A violent explosion at a mobile offshore drilling platform called Deepwater Horizon killed 11 workers, and the platform burned uncontrollably

Fire boats battling the blazing remains of the Deepwater Horizon mobile offshore oil-drilling platform on April 21, 2010. The explosion killed 11 workers and sent millions of gallons of crude oil into the Gulf of Mexico—threatening ecologically fragile coastal regions of Louisiana, Mississippi, Alabama, and Florida. *(USCG)*

until it sank. For the next three months, crude oil spewed from the damaged wellhead on the seafloor, threatening marine life and the ecologically fragile coastal regions of Louisiana, Mississippi, Alabama, and Florida. The multinational oil company BP (British Petroleum) was leasing the drilling platform. On July 15, emergency response workers managed to achieve a static kill on top of the well by pumping mud and cement into the well's containment cap. American officials overseeing response to this major environmental disaster estimated that a total of about 5 million barrels (205 million gallons) of oil had spilled from the ruptured well into the Gulf of Mexico before the capping operations stopped the leakage. By mid-September, a relief well successfully intersected the blown-out oil well some 18,000 feet (5,500 m) below the Gulf of Mexico's seabed. This allowed petroleum engineers to permanently seal the rupture from below. They used the relief well to perform a companion "bottom kill" by injecting mud and cement. The five-month saga represents the world's worst accidental oil spill.

Workers extract crude oil from a normally operating well and ship it to a refinery for processing. Two major techniques for shipping oil are by sea using tankers or barges and overland using a network of pipelines. Both

The USNS *Richard G. Mathiessen* is one of five chartered oil tankers that help transport more than 1.7 billion gallons ($6.4 \times 10^9$ l) of fuel annually for the Department of Defense. *(USN)*

techniques usually transport large quantities of crude oil efficiently each day. However, accidents can occur and result in the spill of crude oil into environmentally sensitive regions. For example, on March 24, 1989, an oil tanker, the *Exxon Valdez,* hit a reef in Prince William Sound, Alaska, and spilled an estimated 10.8 million gallons ($40.9 \times 10^6$ L) of crude oil. Regarded as one of the major human-caused environmental disasters ever, the oil contaminated some 1,300 miles (2,090 km) of ecologically fragile coastline and covered about 11,000 square miles (28,510 $km^2$) of the Pacific Ocean.

The 10 leading crude oil–producing countries in descending order are Saudi Arabia, Russia, the United States, Iran, China, Canada, Mexico, the United Arab Emirates, Kuwait, and Venezuela. The top crude oil–producing states in the United States are Texas, Alaska, California, Louisiana, and New Mexico. More than a fourth of the crude oil produced in the United States is extracted offshore in the Gulf of Mexico. Crude oil is traded in barrels (bbls). Due to density differences after refining and processing, a 42-gallon (159 L) barrel of crude oil typically yields about 45 gallons (170 L) of refined petroleum products.

## PETROLEUM REFINING AND PROCESSING

A modern oil refinery is an amazing factory that takes in crude oil and converts it into a variety of more useful products, including transportation fuels, heating oil, fertilizers, petrochemicals, asphalt, and many other materials. A modern refinery operates 24 hours a day every day of the year. Generally, a refinery will convert a 42-gallon (159 L) barrel of crude oil into 18.56 gallons (70.25 L) of gasoline, 10.31 gallons (39.0 L) of diesel fuel, 1.68 gallons (6.36 L) of heavy fuel oil (residuum), 4.07 gallons (15.41 L) of jet fuel, 1.38 gallons (5.22 L) of other distillates (such as heating oil), 1.72 gallons (6.51 L) of liquefied petroleum gases (LPG), and 7.01 gallons (26.53 L) of other products. These other products include ink, crayons, bubble gum, dishwashing liquids, deodorants, eyeglasses, tires, CDs and DVDs, heart valves, and ammonia.

Each refinery is a unique assembly of processing equipment. However, every petroleum refinery performs three basic steps: separation, conversion, and treatment. All refineries have distillations units that separate crude oil by boiling point. As shown in the illustration, heavy petroleum fractions are on the bottom of the distillation tower, while light fractions occupy the top. This arrangement allows for the separation of various petrochemicals.

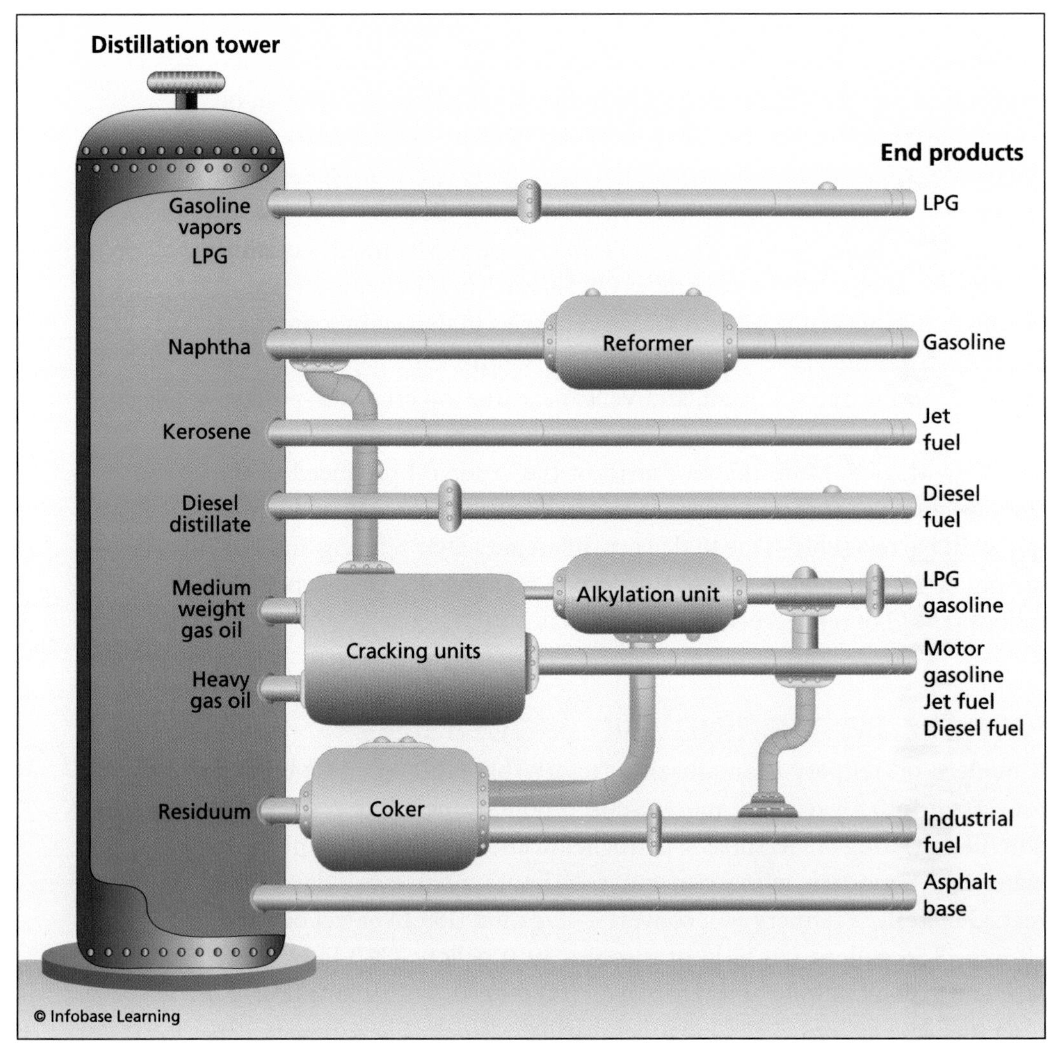


Simplified diagram showing the most important processes in a modern petroleum refinery *(Modeled after DOE)*

Chemists describe petroleum refining as the conversion of the relatively abundant alkanes (found in crude oil) into unsaturated hydrocarbons, aromatic hydrocarbons, and hydrocarbon derivatives. Alkanes (paraffins) are saturated hydrocarbons with the general molecular formula $C_nH_{2n+2}$. Methane ($CH_4$), hexane ($CH_3[CH_2]_4CH_3$), and octane ($CH_3[CH_2]_6CH_3$) are examples of straight-chain alkanes. In the fractional distillation of petroleum, chemist engineers vaporize the input crude oil and then column separate the components according to boiling points. The lower–boiling point constituents, such as naphtha and kerosene, reach the upper portions of the distillation tower, while the higher–boiling point components remain in the lower parts of the tower. Chemical engineers define *naphtha* as the lightest and most volatile fraction of the liquid hydrocarbons in a distillation tower. They use naphtha as the feedstock in the production of high-octane gasoline. The reformer converts naphthas with low octane ratings into gasolines with higher octane ratings.

Gasoline is a hydrocarbon fuel used as the chemical energy source for many internal combustion engines, especially those found in automobiles. Commercial grades of gasoline have molecular compositions ranging from $C_5H_{12}$ to $C_{12}H_{26}$ and boiling points from 104°F to 392°F (40°C to 200°C). On occasion, the air-gasoline mixture in an internal combustion engine prematurely ignites before the spark plug fires. Automotive engineers refer to this undesirable phenomenon as knocking. In the 1920s, chemical engineers created an arbitrary scale for automotive engine performance based on a gasoline mixture's ability to prevent knocking. They discovered that certain blends caused less knocking and assigned these hydrocarbon blends a higher octane rating. Chemical engineers perform a variety of processes and treatments in a modern refinery to produce high-octane gasoline and other desired petroleum products.

They use the process of cracking to break down larger hydrocarbon molecules into smaller size ones. The cracking process allows chemical engineers and chemists to modify natural materials into a variety of new ones more suitable for human needs. In addition to fuels for transportation and heating, crude oil yields a wide variety of petrochemicals that form an integral part of modern living.

The process of alkylation allows chemical engineers to combine certain small hydrocarbon molecules (below the molecular-size range of gasoline) into the large hydrocarbon molecules typically found in gasoline. For example, by reacting isobutylene ($C_4H_8$) with propane ($C_3H_8$), engineers formed highly branched product molecules that were just the proper size for gasoline and were also high in octane number. The hydrocarbon molecule octane has the chemical formula $CH_3(CH_2)_6CH_3$.

Finally, the coker in a petroleum refinery uses heat to transform a portion of the higher molecular weight residuum (residual crude oil) in the distillation tower into lower molecular weight hydrocarbons, such as liquid petroleum gases (LPGs) and gasoline. The remainder of the residuum becomes industrial fuel or asphalt base. This has been a very brief summary of the complex and elegant processing technologies used by chemical engineers in modern petroleum refineries to provide a wide variety of contemporary products that include automotive fuels, pharmaceuticals, plastics, perfumes, and pesticides.

## PETROLEUM AND THE ENVIRONMENT

There is no question that petroleum products drive modern civilization. Petroleum fuels automobiles, trucks, aircraft, and ships. People use petrochemicals at work, in the home, and at play. Nevertheless, petroleum use can also cause undesirable environmental consequences. The combustion of fossil fuels, such as gasoline and diesel fuel, creates atmospheric pollution and increases the carbon dioxide ($CO_2$) burden in the atmosphere. Although significant progress in exhaust pollution control has taken place in the past few decades in the United States, urban areas can still experience severe episodes of smog from automotive emissions.

This prophetic 1971 stamp from Monaco bemoans the growing worldwide threat of oceanic oil pollution. *(Author)*

When petroleum industry geologists explore and drill for oil, they can disturb natural land or marine habitats. Horizontal drilling technologies have reduced the extent of the disturbed areas (called oil production *footprints*), but abandoned or improperly capped wells represent lasting environmental blights.

Oil spills and the blowout of production wells are major, acute threats to the environment—especially when large quantities of crude oil gush out into a fragile environmental region. Marine engineers have designed double-hulled tankers to protect against oil spills during ship collisions or groundings. Tanker accidents and well blowouts represent highly visible episodes of oil pollution that occur in a short period. The local wildlife generally experiences a severe shock, and long-term recovery of the contaminated region often depends on the quality, speed, and extent of the postaccident cleanup activities.

Plus, there are other sources of oil pollution. These include natural seepage from the floor of the ocean, intentional discharges from ships at sea during normal operations, precipitation-induced runoff from roadways and urban areas into bodies of water, groundwater contamination due to seepage from leaky storage tanks, and intentional discharges during war or as a result of terrorist activities. On January 23, 1991, the former dictator of Iraq, Saddam Hussein (1937–2006), intentionally dumped millions of gallons of crude oil directly into the Persian Gulf in an effort to thwart a possible amphibious assault by Coalition Forces during Operation Desert Storm.

The actual amount (percentage) of oil pollution attributed to each of these sources varies considerably from estimate to estimate. However, while tanker accidents and oil well blowouts cause headline-capturing environmental problems, the majority of anthropogenic oil pollution comes not from a few large catastrophes but from an accumulation of thousands (if not millions) of widely dispersed small sources—each of which helps to contribute to the overall problem. Most of these small sources of pollution are directly associated with use of petroleum products by human beings.

Oil or gasoline leaking from storage tanks, including those at neighborhood gas stations, can seep silently into the ground and eventually contaminate local aquifers. Rainwater carries oil and gasoline deposited on highways or spilled at filling stations into local storm sewer systems. The oil-contaminated rainwater typically flows into nearby streams, rivers, and ultimately the ocean. Oil and gasoline pollution from a variety of personal watercraft, including jet skis and motorboats, can directly contaminate lakes, reservoirs, and rivers.

In 2003, the National Research Council (NRC) of the U.S. National Academy of Sciences suggested that about 1.3 million tons of petroleum entered the oceans annually. The following percentages characterize the sources of this estimated annual release: 47.3 percent from natural seepage through the seafloor, 37.9 percent due to discharges from the routine consumption of petroleum (including operational discharges from marine vessels and discharges from sources on land), 11.8 percent due to accidental oil spills from ships, and 3 percent due to oil extraction activities. Oil pollution is detrimental to marine life and aquatic birds. Oil-coated marine organisms and birds become incapacitated and typically perish; the direct ingestion of oil can poison them or other animals feeding upon them. Responsible government agencies must remain vigilant in order to minimize the amount of petroleum released by human actions into coastal waters.

## MERCURY

Mercury (Hg) is the only metal that is liquid at room temperature (68°F [20°C]). Named after the messenger god in Roman mythology, this potentially deadly liquid metal is a potent neurotoxin and is particularly damaging to the development of fetuses, infants, and young children. Mercury is an element, and humans can neither create nor destroy it. It exists in several forms: elemental or metallic mercury, inorganic mercury compounds, and organic mercury compounds.

In about 1600 B.C.E., the metal became known to ancient peoples around the Mediterranean Sea. Mercury soon gained wide use because of its ability to dissolve gold and silver. Gold- and silver-bearing ores were crushed and treated with mercury. As the liquid mercury formed a solid amalgam, it extracted the precious metals from the host ore. (An amalgam is an alloy containing mercury.) Heat treatment of this particular amalgam would drive off the mercury, leaving behind the sought-after gold and silver.

With a melting point of –37.9°F (–38.8°C [234.3 K]), mercury (Hg) is a liquid under room temperature conditions. In its gaseous state, mercury is a colorless vapor. The metal gallium (Ga) has a melting point of 85.6°F (29.8°C [302.9 K]), which means this metal is a liquid at just above room temperature, so a small piece of it would melt in the palm of a person's hand. Similarly, cesium (Cs) is a metal that transitions to the liquid state at just above nominal room temperature conditions. Cesium is an alkali metal with a melting point of 83.2°F (28.4°C [301.6 K]). Bromine (Br), with a melting point of 19.0°F (–7.2°C [266 K]), is the only nonmetallic element that is liquid at room temperature.

Mercury (Hg) has an atomic number of 80 and an atomic weight of 200.59. The Latin name for mercury is hydragyrum. Sometimes called quicksilver, mercury is usually not found free in nature. This silvery-white liquid metal has a density of 845.7 $lbm/ft^3$ (13.546 $g/cm^3$) at room temperature and breaks up into tiny drops or beads when spilled. It is generally obtained from the mineral cinnabar (HgS).

The liquid metal readily forms alloys (called amalgams) with gold, silver, zinc, and cadmium. Mercury is toxic and a major environmental pollutant. Today, mercury is commonly used in scientific instruments such as thermometers and barometers, personal electronics, fluorescent lamps, and mercury vapor street lights. Mercuric chloride ($HgCl_2$) is a very poisonous salt that was once used by physicians to disinfect wounds. Up until the end of the 19th century, people also used mercurous chloride

($Hg_2Cl_2$), or calomel, as an antiseptic and an insecticide. Physicians used this colorless, white, or brown tasteless compound to treat sexually transmitted diseases. Even in the 21st century, people have suffered from mercury poisoning by using facial crèmes containing calomel. (Such crèmes are officially banned for use in the United States by the Food and Drug Administration [FDA].)

Mercuric sulfide (HgS) is used to make vermilion, a red paint pigment. The compound mercuric oxide (HgO) is used in the manufacture of mercury batteries. Some paint manufacturers have routinely added phenylmercuric acetate to interior latex (water-based) paint as a fungicide and bactericide to prolong the paint's shelf-life. Before 1990, the U.S. Environmental Protection Agency (EPA) had permitted interior latex paint to contain less than or equal to 300 parts per million (ppm) elemental mercury and exterior latex paint to contain less than or equal to 2,000 ppm. Interior latex paint is now mercury free, although exterior latex paint may still contain mercury.

Mercury is a highly toxic chemical element found naturally in the environment as a result of volcanic eruptions and emissions from the oceans. It also a toxic contaminant frequently introduced into the environment by human activities. The major sources of mercury emissions in the United States are coal-fired power plants, waste incinerators, and metallurgy/mining operations, particularly gold mining. The improper disposal of fluorescent lightbulbs and personal electronics containing mercury represents a growing source of mercury in the environment. When crushed and broken in a landfill, mercury contained in such items eventually leaks and can escape as a vapor into the atmosphere.

What is especially troublesome about mercury is that it exists in a variety of chemical and physical forms and can change from one form to another. Some forms of mercury are much more toxic than others. Environmental scientists know that, depending on the form, mercury released in the atmosphere can travel short or long distances—even all the way around the world. Mercury can also cycle between the atmosphere and Earth's surface. Unless it is contained, mercury in the surface environment can be readmitted into the atmosphere. Estimates made by the Department of State in 2007 suggest that more than half of all mercury deposition in the United States comes from sources outside national boundaries. For example, the combustion of coal, linked to the economic growth of Southeast Asia, is the fastest-growing source of mercury emissions affecting the air quality and environment of the United States.

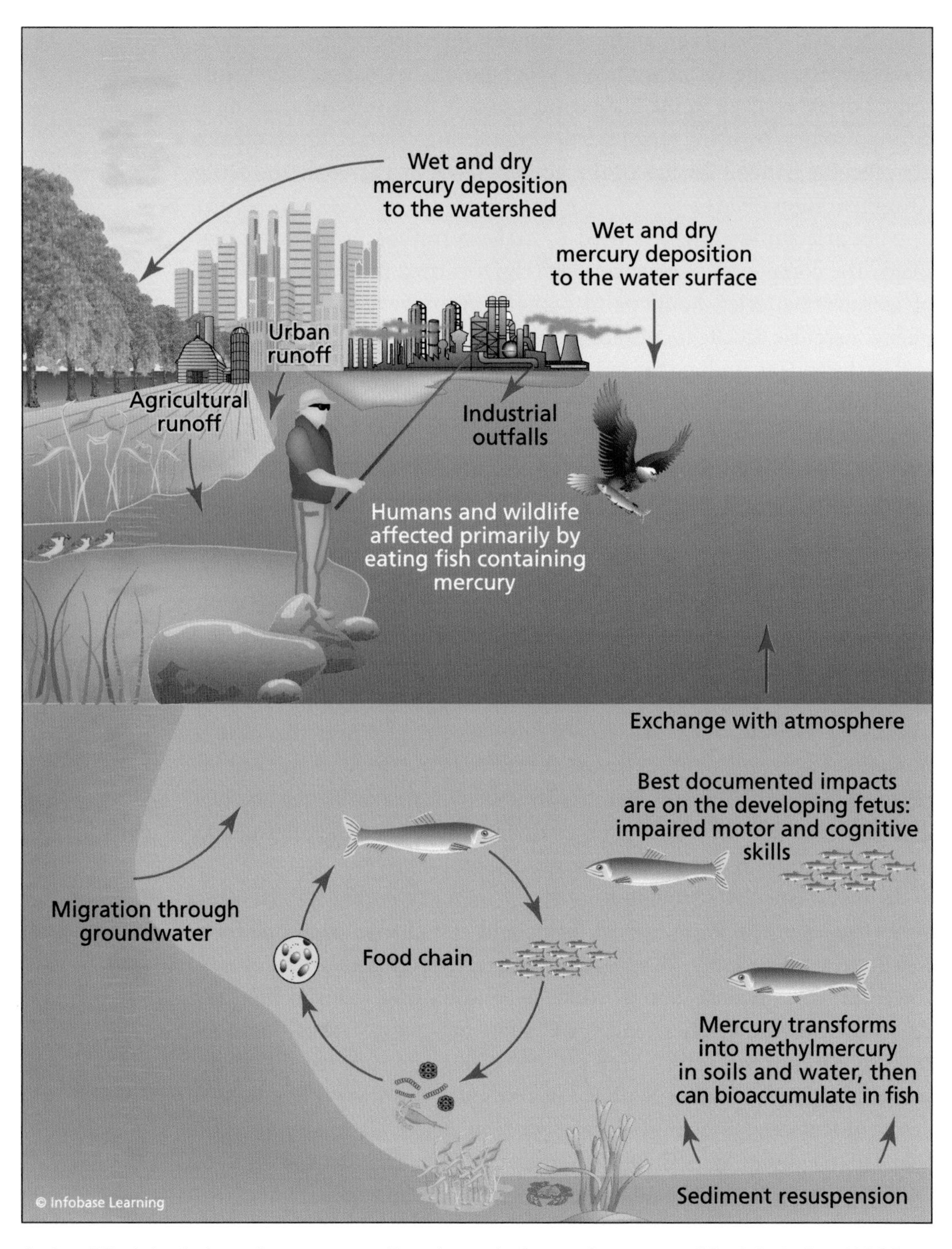


A simplified depiction of mercury cycling through the environment *(Modeled after USGS and EPA)*

Physicians call mercury poisoning hydrargyria. Mercury can enter the human body through the respiratory tract, the digestive tract, or directly through the skin. It accumulates in the body and eventually causes severe illness or death. The toxic effects of mercury depend on its chemical form and the pathway of exposure. Methylmercury ($CH_3Hg$) is the most toxic form. Exposure to methylmercury usually happens by ingestion, such as by eating mercury-contaminated fish. The human body absorbs methylmercury more readily and excretes it more slowly than other forms of mercury.

Vaporous elemental mercury, symbolized as $H^0$ in environmental studies, causes tremors, impairs hearing and speech, and may promote excitability if the vapors are inhaled over a prolonged period of time. Mercury (II) nitrate, $Hg(NO_3)_2$, was a soluble, white, crystalline compound of mercury that hatters used to treat the felt for making hats. As a consequence, many hatters in the 19th century suffered from mercury poisoning, giving rise to the expression "Mad as a hatter." The prevalence of hydrargyria among hatters may have inspired Lewis Carroll (1832–98) to include The Hatter, now more commonly referred to as The Mad Hatter, in his novel *Alice's Adventures in Wonderland.*

Mercury's biological hazard is not uniform. If a person accidently ingests elemental liquid mercury (perhaps from a broken oral thermometer), it may pass through the individual's digestive system without causing damage. This happens because the metal is absorbed very slowly by the body.

Mercury in the atmosphere eventually settles into water or else falls on land and then may get carried by rain into bodies of water. Once deposited in water, certain microorganisms can transform mercury into methylmercury, a highly toxic compound that bioaccumulates in fish and shellfish. Methylmercury also builds up in animals, including humans, that eat mercury-contaminated fish and shellfish. Because of the peculiarities of the bioaccumulation and biomagnification processes, methylmercury builds up more in some types of fish and shellfish than others. The levels of methylmercury in fish and shellfish depend on what they eat, how long they live, and where they reside in the local food chain.

Scientists define biomagnification as the bioaccumulation of a substance up the food chain by the transfer of residues of the substance present in smaller organisms that serve as food for larger organisms in the food chain. Generally, the concentration of the substance in question is higher in organisms farther up the food chain. Biomagnification causes higher concentrations than anticipated if contaminated water were the only exposure mechanism. Scientists define bioconcentration as the accumulation of a substance only through contact with water. Because of bioaccumulation, both birds and mammals, including humans, that eat methylmercury-contaminated fish or shellfish are more exposed to mercury than other animals in the aquatic ecosystems. At higher levels of exposure, methylmercury's harmful effects on birds and mammals includes impaired reproduction, slower growth and physical development, abnormal behavior, and possibly death.

For people of all ages, high levels of mercury exposure can injure the brain, heart, kidneys, lungs, and immune system. According to scientists at the EPA, high levels of methylmercury in the bloodstream of unborn babies or young children could harm the developing nervous system, making the child less capable of thinking and learning. From the 1930s to the 1960s, a chemical factory near Minamata Bay, Japan, routinely discharged its wastes containing methylmercury into the bay. In the 1950s, thousands of local people who had been eating fish and shellfish from the bay began showing signs of neurotoxin poisoning, later named Minamata disease. Symptoms included loss of muscle control, slurred speech, convulsions, sudden fits of laughter or other crazed mental states, coma, and death.

CHATPER 9

# Volcanoes, Hot Springs, and Geysers

**This chapter examines some of the more interesting natural phenomena** that involve the flow of heated liquids. These phenomena include lava flowing from volcanoes; the heated water, spectacular geysers, and bubbling mudpots (also spelled mud pots) associated with hot springs; and hydrothermal vents found on the ocean floor. Space exploration indicates similar geothermal phenomena exist on other worlds in the solar system. The chapter concludes with a discussion of geothermal energy systems.

## RESTLESS EARTH

Earth is a restless planet that evolved during the past 4.6 billion years, transitioning from a totally molten world into a habitable planet possessing a solid outer skin. The surface of humans' home planet continues to be influenced by the Sun, gravitational forces, complex interactions with the hydrosphere and atmosphere, and processes emanating from deep within Earth's core. On very short time scales, people appear to be standing on solid ground, or terra firma. Despite such stable appearances, Earth continuously experiences many geophysical processes that transform and sculpt its surface. These planet-changing processes can be quite dramatic and lead to loss of life as well as involve extensive property damage. Sometimes, these changes happen slowly, such as surface subsidence due to aquifer depletion); other times, the changes are unpredictable and rapid,

A geologist measures the height of a lava fountain during the 1983 eruption of the Kilauea Volcano at Hawaii Volcanoes National Park. *(USGS)*

such as those brought about by earthquakes, volcanic eruptions, and landslides.

Earth's surface and interior are major components of an interconnected, highly dynamic physical system scientists call the solid Earth. Violent events, such as earthquakes, volcanic eruptions, landslides, and floods, reshape the surface and pose significant hazards. Solid-earth scientists attempt to measure both the slow and fast deformations of the planet's surface in order to improve their overall understanding of the more dominant geological processes. This improved knowledge will help them optimize responses to natural hazards and identify potential risk areas.

Earth has an equatorial radius of 3,963 mi (6,378 km). Starting at the surface and descending (the direction of increasing pressure, temperature, and density), the planet consists of three distinctive layers: crust, mantle, and core. Scientists find it convenient to divide these layers according to their respective densities. The crust forms the rigid but thin, rocky outer skin of the planet. It is about 31 miles (50 km) thick or less and contains low-density rock materials, such as granite and basalt. Below the crust, there is a thick layer, about 1,800 miles (2,900 km) thick, called the mantle. The mantle consists of hot, rocky material of modest density and contains minerals mixed with silicon, oxygen, and other elements. Below the mantle is the high-density spherical core, consisting of an outer (liquid) region, about 1,400 miles (2,260 km) thick, and an inner (solid) region, about 727 miles (1,170 km) in radius—each of which contains metals, especially iron and nickel.

The outermost layer of the mantle is solid and, along with the crust, forms a region scientists define as the lithosphere. The lithosphere is about 93 to 124 miles (150 to 200 km) thick under the continents but thinner under the oceans. The crust and upper mantle are subjected to the dynamic forces produced by mantle convection, which causes segments of the shells and materials at the top to break up into plates. The term *tectonics* refers to any planetary surface changes that result from the compression (squeezing), stretching (tension), or other forces acting upon the lithosphere. Scientists define plate tectonics as the geological process whereby plates move under, over, or around each other in the lithosphere. When two plates collide, surface compression occurs, and mountain building takes place.

## VOLCANOES

A volcano is a vent or opening at the Earth's surface through which magma (molten rock) and associated gases erupt. Scientists also use the term *volcano* to describe the cone that has been built up by effusive (over-the-top) and explosive eruptions.

Throughout history, volcanoes have represented some of Earth's most powerful and destructive natural forces. On August 24, 79 C.E., the Roman cities of Pompeii and Herculaneum were destroyed when the long-dormant Mount Vesuvius volcano suddenly exploded. Within just a few hours, hot volcanic ash and dust buried the two cities, killing thousands of residents.

The accompanying illustration describes the major components of a typical volcano. Far below the surface, magma accumulates in a hollow portion of the lower strata called a magma chamber. Streams of magma then rise from this chamber along a central vent toward an opening on the surface. Scientists call the raised opening that emits magma the cone. When magma erupts onto the surface and flows out of the cone, they refer to the hot, less viscous flowing material as *lava*. An erupting volcano may also emit a stream of rock and ash, which scientists call tephra.

At temperatures ranging between 1,292°F (700°C) and 2,192°F (1,200°C), lava is the hottest substance found naturally on Earth's surface. Significantly more viscous than water, lava can flow for a great distance before cooling. Once cooled, lava solidifies into igneous rock. Geologists use the term *lava flow* for the outpouring of molten rock associated with a nonexplosive volcanic eruption. They refer to very low-density, highly porous solidified lava as pumice. Because of its numerous entrapped gas bubbles and air chambers, pumice has a sufficiently low density to float on water. Scientists often find pumice being carried by ocean currents far away from volcanic islands.

In native Hawaiian mythology, Pele is the volcano goddess associated with fire, violence, and lightning. The Hawaiian language is very rich in

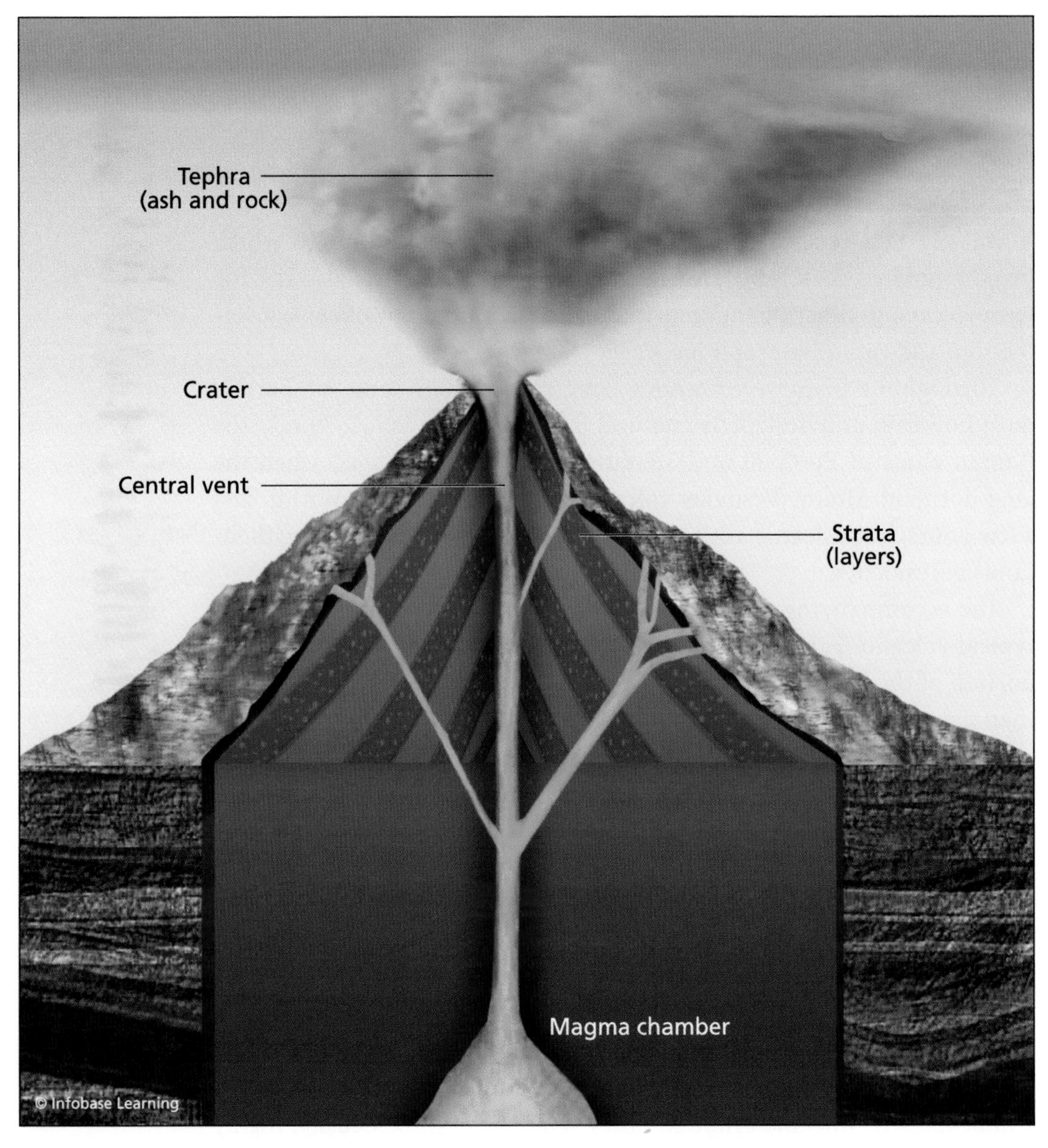


This illustration provides a cutaway view of a typical volcano. *(USGS)*

terms describing volcanic phenomena and lava flows. Two examples are provided here. The Hawaiian term *'a 'a* (pronounced "ah-ah") refers to lava flows that have a rough surface that consists of broken blocks of lava. The Hawaiian term *pahoehoe* denotes basaltic lava flows that have a smooth, ropy (hummocky) surface. Geologists use the term *hummock* to identify a small mound above ground. According to geologists, a pahoehoe flow usually advances as a series of small toes or lobes that continuously break out from a cooled crust. The result is a relatively smooth surface texture that exhibits a wide variety of bizarre shapes. *'A 'a* and pahoehoe are the two most common lava flows associated with Hawaiian Island–type volcanoes.

Magma is molten or partially molten rock that lies beneath Earth's surface. Magma consists of a liquid portion, a solid portion, xenoliths, and dissolved gases. Scientists sometimes refer to the liquid portion of magma as *melt.* The solid portion of magma consists of minerals that have crystallized directly from the melt. Xenoliths (or inclusions) are adjacent solid rock materials incorporated into the magma as it flows within magma chamber or up through various conduits, including the central vent. When magma erupts onto the planet's surface, scientists call the hot flowing substance lava.

The viscosity of the magma determines the nature of the volcanic eruption. Some eruptions are explosive, while others involve a relatively gentle flow of lava out of the cone. Entrained gases cannot easily escape thick, highly viscous magma. Pressure builds up in such magmas until there is a violent explosion that liberates the accumulated gases. During an explosive volcanic eruption, the emerging magma bursts into the atmosphere and disintegrates into pieces of hot rocks and ash called tephra. Tephra ranges in size from car-size boulders to tiny particles of ash. If the magma is less viscous, it flows more easily out of the volcano as lava, and any entrained gases easily escape into the atmosphere.

Explosive volcanic eruptions are destructive and deadly. During such eruptions, the side or top of the volcano often disappears as clouds of hot tephra blast out. Scientists define a pyroclastic flow as a ground-hugging avalanche of rock fragments, pumice, hot ash, and escaping volcanic gases that rushes down the side of a volcano at speeds of 60 miles per hour (about 100 km/hr) or more. The temperature within a pyroclastic flow can exceed 932°F (500°C).

On Earth, volcanoes occur because the crust is broken into rigid plates that float on the softer layer of rock within the mantle. As the plates move, they push together or pull apart. Most terrestrial volcanoes occur near the edges of these tectonic plates. When tectonic plates collide (push together), one plate slides beneath the other. Geologists refer to such a collision zone as the subduction zone. When the descending plate plunges deep enough into the mantle, some of the rock on the overlying plate melts, forming magma that can ascend and eventually erupt as lava. Geologists call regions where tectonic plates are drifting apart rift zones. As plates move apart in a rift zone, magma can also rise to the surface and cause a volcanic eruption. Finally, there are regions in the middle of tectonic plates called hot spots where magma melts through cracks and crevices within the plate and eventually causes a volcanic eruption.

## EXTRATERRESTRIAL VOLCANISM

Space exploration has allowed scientists to examine other worlds in the solar system for signs of volcanism and related geothermal phenomena. Decades of scientific exploration by an armada of robot spacecraft have provided geologists with amazing new insights. In addition, the astronaut Harrison H. Schmitt (1935– ), a professional geologist, walked on the Moon in December 1972 during NASA's *Apollo 17* lunar landing.

Scientists now recognize that impact cratering represents the most important geological process on both the Moon and the planet Mercury, although the interiors of both worlds might have been hot enough in the very distant past to permit tectonics and some volcanism. Today, both relatively small worlds are geologically dead. Space probes have revealed dramatic evidence of extremely active volcanism on ancient Mars. One startling example is Olympus Mons, a towering shield volcano, the largest in the solar system, that measures 374 miles (602 km) across at the base and rises approximately 16 miles (26 km) above the surface of the Red Planet. Radar-imaging data suggest that cloud-enshrouded Venus has experienced recent volcanism and is probably geologically active today. The Jovian moon Io is the most volcanically active body in the solar system. Finally, Enceladus, a medium-size moon of Saturn, appears to possess a subsurface reservoir of liquid water that vents into space like a fountain during times of geological activity.

Repeated eruptions allow a volcano to grow and form a distinctive shape based upon the molten materials involved in the eruption. Geologists divide terrestrial volcanoes into three general types: stratovolcanoes, cinder volcanoes, and shield volcanoes. Resembling a layer cake with frosting, a stratovolcano forms a symmetrical cone with steep sides as each eruption causes lava and tephra to accumulate in strata, or layers. A cinder volcano is a very small, cone-shaped volcano. When erupting lava blasts into the atmosphere and disintegrates into small pieces, the pieces cool and harden into cinders. The cinders then fall back to Earth and accumulate around the volcano's vent, creating the geologic structure known as a cinder volcano. Finally, a shield volcano forms when successive nonexplosive (effusive) eruptions allow lava to smoothly flow from the cone, spread out across the surface, and produce a volcano with broad, gently sloping sides.

## HOT SPRINGS, GEYSERS, FUMAROLES, AND MUDPOTS

When established in 1872, Yellowstone National Park became the first American national park. Located in the states of Wyoming, Montana, and Idaho, the park is home to a large variety of wildlife, including grizzly bears, wolves, bison, and elk. Preserved within this beautiful park is a collection of extraordinary geothermal features such as geysers, hot springs, fumaroles, and mudpots. Yellowstone's geothermal features occur in the only essentially undisturbed geyser basin remaining in the world. In Iceland and New Zealand, for example, human-made geothermal drill holes and wells have significantly reduced hot spring discharges and geyser activities.

More than 75 percent of the world's geysers are found in Yellowstone Park. The park's most famous geyser is appropriately named Old Faithful. According to National Park Service (NPS) records, Old Faithful has been Yellowstone's most dependable geyser, erupting to a height of more than 100 feet (30.5 m) with an average time between eruptions ranging from 66 to 80 minutes.

Essential to the existence of geothermal features such as hot springs, geysers, fumaroles, and mudpots is the presence of a shallow magma chamber beneath the surface. The available groundwater and the physical characteristics of the ambient rock materials establish the natural plumb-

Old Faithful Geyser erupts in winter at Yellowstone National Park. *(NPS)*

ing system necessary for the operation of specific types of geothermal systems. For example, in the high mountains surrounding the Yellowstone Plateau, water that falls as rain or snow eventually percolates through layers of porous rock. After descending to a depth of nearly 10,000 feet (3,050 m), the cold groundwater encounters the hot rocks associated with the shallow magma chamber. Upon contact, the water's temperature rises well above the normal boiling point (namely, 212°F [100°C] at one atmosphere pressure) and becomes superheated.

Due to the high lithostatic pressure at this depth, the superheated water remains in the liquid state even though its temperature exceeds 400°F (204°C). The less dense superheated water then begins its slow, arduous journey toward the surface by wandering through cracks, fissures, and weak areas within Earth's crust. At Yellowstone, ancient rhyolitic lava flows have created a natural plumbing system in the crust that encourages the formation of hot springs and, under special circumstances, geysers.

As the superheated water rises up through this complex natural plumbing system and nears the surface, the immense pressure exerted on the water deep underground near the magma chamber begins to drop. If released in a slow, steady manner, the captured geothermal energy in the heated water feeds one of the many colorful hot springs found throughout Yellowstone Park.

Hot springs are found throughout the globe in both volcanic and nonvolcanic areas. In nonvolcanic areas, groundwater percolates deep into the crust and extracts geothermal energy from the surrounding rocks; in volcanic areas such as Yellowstone, groundwater encounters a shallow magma chamber just beneath the surface.

Sprinkled amid the numerous hot springs of Yellowstone National Park are geysers, the rarest and most dramatic fountains found in nature. *Geyser* is an Icelandic word meaning "to gush or rush forth." Fine-grained, light-colored, acidic volcanic rock called rhyolite is essential to the formation of geysers. Rhyolite contains an abundance of silica (silicon dioxide [$SiO_2$]). As hot water travels up through the rhyolite, some of the silica is dissolved. At the surface, the silica-laden water forms an interesting rock geologists call geyserite (or sinter). Under certain conditions, geyserite accumulates on the walls of the natural underground plumbing system of a hot spring and creates a pressure-tight seal. Hot water collects in this confined space until thermodynamic conditions trigger a geyser's episodic eruption.

It is the presence of one or more such constrictions just below the surface that distinguishes a geyser from the more commonly encountered hot spring. In a geyser, expanding steam bubbles accumulate in a confined space behind a constriction. The buildup of steam eventually forces overlying water up through the constricted natural conduit and out onto the ground. This rapid loss of water reduces the pressure within the confined space. Suddenly, superheated water deep within the conduit flashes to steam and erupts violently into the atmosphere, creating a spectacular natural fountain of scalding water and steam. Engineers familiar with thermodynamics view an erupting geyser as a naturally occurring steam explosion during which superheated water experiences a rapid loss of pressure and a rapid expansion into the surrounding atmosphere. The sudden rupture of a human-made pipe carrying a liquid at high pressure produces a similar geyser effect.

Fumaroles are steam vents that occur along tiny fissures and surface cracks in volcanic areas. Fumaroles are essentially hot springs that have very little groundwater. As a result, any water boils away and vents as steam before it can reach the surface. In addition to steam, fumaroles also emit volcanic gases such as sulfur dioxide, hydrogen sulfide, and carbon dioxide.

Finally, mudpots form in certain geothermal areas where the supply of water is limited and hydrogen sulfide gas is also present. These circumstances enable the generation of sulfuric acid, which dissolves surrounding rock into fine particles of silica and clay. The clay and silica particles then mix with available hot water to form bubbling mudpots. Yellowstone Park has numerous examples of these colorful geothermal features, which are both noisy and smelly. The foul smell of a mudpot comes from the presence of hydrogen sulfide ($H_2S$) gas, a colorless, poisonous gas with the odor of rotten eggs.

## HYDROTHERMAL VENTS

In 1977, marine scientists discovered an underwater (submarine) hydrothermal vent along the Galápagos Rift in the Pacific Ocean. Many more of these interesting geothermal phenomena have since been discovered on the floor of both the Atlantic and Pacific Oceans.

Scientists have detected the most numerous and spectacular hydrothermal vents along the Earth's midocean ridges. The heat source for these submarine hot springs (or underwater geysers) is the magma beneath the volcanic ridge system. Marine geologists suggest that an underwater hydrothermal vent is basically a geyser on the seafloor. In

Black smoker hydrothermal vent at a mid-ocean ridge in the Atlantic Ocean *(NOAA; OAR/National Undersea Research Program [NURP])*

areas where tectonic plates are moving apart, seawater seeps into cracks and fissures in the ocean floor and is then heated by contact with a subsurface magma chamber. The hydrothermal vent spews out superheated water at up to about 752°F [400°C]) that is rich in minerals. Because of hydrostatic pressure, geothermal activity beneath 6,560 feet (2,000 m) to 16,400 feet (5,000 m) of seawater is considerably different than on land. The hot fluid flowing through a submarine hydrothermal vent is very corrosive and capable of dissolving the surrounding basaltic rock. As the

## CHEMOSYNTHESIS

The vast majority of life on Earth depends on photosynthesis, the amazing process by which green plants, algae, and certain microorganisms use the energy of sunlight to convert carbon dioxide into complex organic molecules, especially sugars. Equally interesting is the fact that at hydrothermal vents in deep portions of Earth's oceans, unique ecosystems have evolved in the absence of sunlight. These ecosystems depend upon chemosynthesis, the process by which microorganisms liberate energy by mediating chemical reactions. In the majority of hydrothermal vent environments, the energy supply for chemosynthesis comes from oxidation reactions of inorganic molecules such as hydrogen sulfide. At other sites, the chemosynthesis of methane ($CH_4$) from carbon dioxide ($CO_2$) and hydrogen drives the production of microscopic life. Many of the chemosynthetic microorganisms dwelling around a hydrothermal vent are consumed by other deep-ocean marine organisms. Where chemosynthetic microbes form a mat covering the seafloor around vents, grazers such as snails, limpets, and worms thrive by eating this mat. Then, predators such as crabs consume the grazers.

The tubeworm is one of the more fascinating creatures found in the web of life around a hydrothermal vent. Reaching up to eight feet (2.4 m) in length, the eyeless and mouthless creature uses its gill to extract oxygen, sulfur, and other substances from the water. In a classic symbiotic relationship, the tubeworm provides a stable biological niche for the billions of bacteria that inhabit its interior, while the bacteria process the chemicals that emerge from the vent into food for the tubeworm. The bacteria further benefit from their relationship with the tubeworm because the host worm delivers blood-containing hemoglobin, a biological fluid that helps the bacteria break down the ingested sulfides.

Scientists suggest that chemosynthesis may also be taking place on other worlds in the solar system. Europa, an ice-covered major moon of Jupiter, is a leading candidate. Long suspected of possessing a large, liquid water ocean buoyant, heated liquid rises rapidly then falls back to the surface of the ocean floor, it leaches out metals and other materials from the surrounding rock in the midocean ridge.

Scientists refer to the most spectacular type of hydrothermal vent as a black smoker. The so-called black smoke gushes from a chimneylike

**Artist's rendering of a future NASA robot exploring postulated hydrothermal vents on the floor of a liquid water ocean beneath Europa's icy surface** ***(NASA/JPL)***

beneath its frozen surface, Europa is subject to tidal heating due to Jupiter's gravitational influence. Such tidal heating could readily promote the formation of hydrothermal vents similar to those found on the floor of Earth's oceans. Given hydrothermal vents, astrobiologists postulate that alien life forms on Europa might form ecological systems based on chemosynthesis.

structure on the seafloor and consists of tiny metallic sulfide particles that precipitate out of the hot vent fluid as it mixes with cold seawater. Black smokers are examples of focused hydrothermal vents, a geothermal phenomenon in which almost all the venting fluid flows out of one small pipe-like conduit. Scientists call submarine hydrothermal vents that are rich in barium, calcium, and silicon white smokers.

## GEOTHERMAL ENERGY

The term *geothermal* refers to the heat associated with Earth's interior. Scientists regard the total amount of heat available within the Earth as enormous, so energy systems that use geothermal energy are treated as renewable energy systems.

Some applications of geothermal energy take advantage of shallow pockets of magma that heat groundwater. Water at elevated temperatures from hot springs has been used since ancient times for bathing, heating, and food preparation. Similarly, modern direct-use geothermal energy systems and district heating systems employ hot water from springs near the surface. Bathing in hot springs remains very popular for many people. Where geographically feasible, engineers pipe the water from hot springs directly into buildings for space heat and into industrial facilities for use as process heat. A modern geothermal energy district heating system provides heat for 95 percent of the buildings in the city of Reykjavik, Iceland. The most common industrial application of geothermal energy is dehydration, the drying of fruit and vegetable products.

Geothermal energy can also be used for electric power generation. The United States now generates more geothermal electricity than any other country in the world, but the amount produced is less than 0.5 percent of all the electricity generated in the United States. The vast majority of geothermal electricity produced in the United States is generated by power plants in California.

There are three basic types of geothermal power plant systems used to convert hydrothermal fluids into electricity: the dry steam power plant, the flash steam power plant, and the binary-cycle power plant. This section briefly describes the features and operational principles of each system. The type of conversion system depends on the thermodynamic state of the naturally heated working fluid (that is, water or steam)—especially the fluid's temperature as it leaves the geothermal source and enters the power generating equipment.

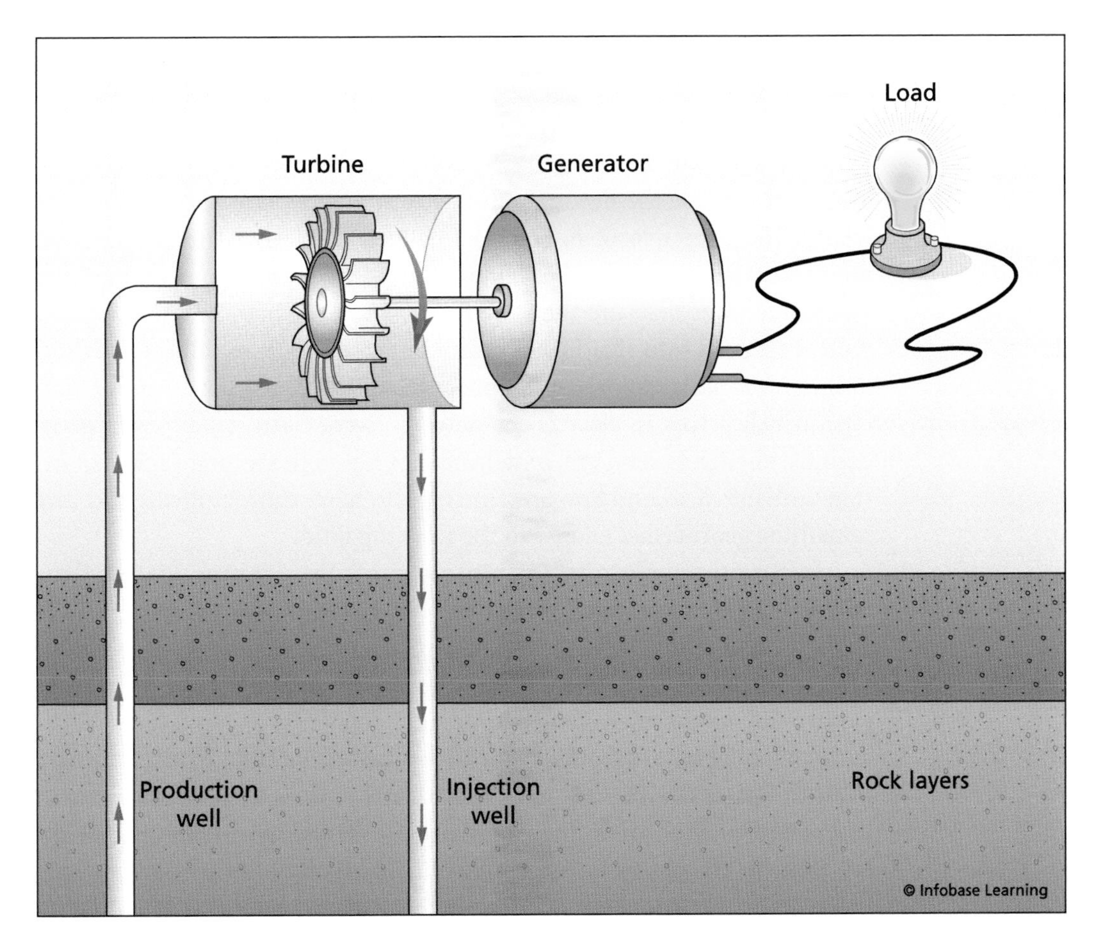


Basic components of a dry steam power plant—the well provides geothermal steam to spin the electric power-generating turbine. *(DOE)*

Dry steam power plants were the first type used for geothermal electric power generation. The first plant appeared in 1904 in Lardarello, Italy. The operational principle remains technically sound, and the approach to electric power generation remains effective—if suitable sources of good quality, dry geothermal steam are available.

This type of plant uses hydrothermal fluids that are primarily steam. Wells take the steam from the geothermal source and feed it directly into a turbine, which drives the generator that produces electricity. The use of naturally heated steam in this type of power plant eliminates the need to burn fossil fuels to generate electricity. Dry steam power plants are used at The Geysers in northern California, the world's largest single source of geothermal power. The generating plants at The Geysers emit only spent (used) steam and very minor amounts of the gases normally associated with volcanic regions.

If the naturally heated geothermal fluid is hotter than 360°F (182°C), engineers can use the fluid in a flash steam power plant to generate electricity. The principle of operation involves extracting the superheated fluid from the ground and spraying it into a large tank that is maintained at a much lower pressure than the fluid. When the high-pressure, heated fluid is sprayed into this low-pressure tank, some (if not all) of the fluid flashes, or rapidly vaporizes. Engineers then use the vapor to spin a turbine, which drives an electric generator. In some flash steam power plant designs, engineers send any remaining hot liquid in the first rapid expansion tank into a second low-pressure tank to more thoroughly extract any remaining geothermal energy in the working fluid.

Most geothermal areas contain moderate temperature (cooler than 360°F [182°C]) water. Such moderate temperature water is not very suitable for use in either dry steam power plants or flash steam power plants, so engineers designed a binary-cycle power plant to more efficiently extract the geothermal energy from such geothermal sites. In a binary-cycle power plant, a secondary (or binary) working fluid extracts energy from the naturally heated water that then becomes a vapor that travels in a closed cycle through the turbine that spins the generator. The working fluid in the secondary (or binary), closed-cycle loop has favorable thermodynamic properties that allow it to readily vaporize at moderate temperatures, pass through the turbine, and then cool back into a liquid by passing through an air-cooled condenser (heat rejection system). The hydrocarbon butane ($C_4H_{10}$) is an example of a candidate binary working fluid for this type of geothermal power system. By design, the secondary fluid circulates in a closed loop and does not get released into the atmosphere. As the moderate temperature hot water passes through the binary-cycle heat exchanger, it yields much of its energy to the binary-cycle working fluid and then travels as cooler water back to the underground geothermal reservoir.

# CHAPTER 10

# Very Cold Liquids

This chapter discusses cryogenic fluids, their physical properties, and their applications. While people normally think about oxygen, nitrogen, hydrogen, and helium as gases, at very low temperatures, these gases become cryogenic liquids with special characteristics and interesting properties.

## CRYOGENICS

Cryogenics is the branch of science that deals with very low temperatures, applications of these very low temperatures, and methods of producing them. Engineers in this field are usually concerned with practical problems, such as producing, transporting, and storing the large quantities of liquid oxygen or liquid hydrogen used for chemical rocket propellants. Engineers are also asked to design the cold-resistant equipment that allows physicists to investigate some basic properties of matter at extremely low temperatures. At extremely low, or cryogenic, temperatures, traditional research materials become very brittle, so scientists must use special equipment when they explore temperature conditions at the threshold of absolute zero.

For engineers, a cryogenic temperature is typically one below –238°F (–150°C [123 K]). However, scientists often use the boiling point of liquid nitrogen at atmospheric pressure—namely, –319°F (–195°C [77.4 K])—

as the cryogenic temperature threshold. In extremely low-temperature research, other scientists often treat the temperature associated with the boiling point of liquid helium—namely, −452.2°F (−269°C [4.2 K])—as the threshold of the cryogenic temperature regime. Liquid helium is the most important cryogenic fluid, or cryogen, in science.

Cryogenics emerged as an important scientific field during the last two decades of the 19th century. In 1883, the Polish scientists Zygmunt Florenty Wróblewski (1845–88) and Karol Stanislaw Olszewski (1846–1915) investigated the properties of liquefied air at the Jagiellonian University in Kraków and became the first researchers to produce scientifically significant quantities of liquefied oxygen and liquefied nitrogen.

Elsewhere in Europe, the Scottish scientist Sir James Dewar (1842–1923) specialized in the study of low-temperature phenomena and invented the double-walled vacuum flask that now carries his name. In 1892, Dewar began using his newly invented vacuum flask device to store liquefied gases at very low temperatures. On May 10, 1898, he used the Dewar flask and innovative regenerative cooling techniques to become the first scientist to successfully liquefy hydrogen.

The Dewar flask, or dewar, is a double-walled container with the interspace (annular space) evacuated of gas (air) to prevent the contents from gaining or losing heat by conduction or convection. The walls are also silvered to minimize radiation heat transfer. Aerospace engineers use large modern versions of this device to store a space launch vehicle's cryogenic propellants, such as liquid hydrogen and liquid oxygen (LOX).

The German engineer Carl von Linde (1842–1934) pioneered refrigeration technology toward the end of the 19th century. His efforts included developing and patenting a process for the liquefaction of atmospheric air. By 1905, Linde had perfected technologies to extract large quantities of pure oxygen and pure nitrogen from liquid air. His efforts made oxygen and nitrogen available for use in science, medicine, engineering, and industry. One consequence of Linde's work was the availability of liquid oxygen for use as a cryogenic propellant in modern liquid-fueled rocket engines.

As a result of the pioneering efforts of Dewar, Linde, and other low-temperature researchers, all temperatures down to −434.2°F (−259°C [14 K]) were attainable in laboratory environments by the beginning of the 20th century. Despite this amazing progress, one elusive goal remained. Although many scientists tried, no one had yet achieved the production of liquid helium. Then, in 1908, the Dutch physicist Heike Kamerlingh Onnes (1853–1926) developed techniques to investigate temperatures

## ABSOLUTE ZERO

Scientists define absolute zero as the temperature at which molecular motion ceases and an object has no thermal energy. In 1848, the Scottish physicist Lord Kelvin (William Thomson) (1824–1907) became the first scientist to formally suggest its existence. Based on the well-proven laws of thermodynamics, absolute zero is the lowest possible temperature and by international scientific agreement represents 0 K (–273.15°C), or 0 R (–459.67°F).

At absolute zero, atoms and molecules are at their lowest but finite energy state. In practice, scientists recognize that it is impossible to achieve absolute zero, because the input power requirements to obtain this theoretical temperature limit would approach infinity. However, research scientists have achieved temperatures that lie within a few billionths of a degree above absolute zero. Extremely low temperature researchers speak in terms of millikelvin (mK) ($10^{-3}$ K),

*(continues)*

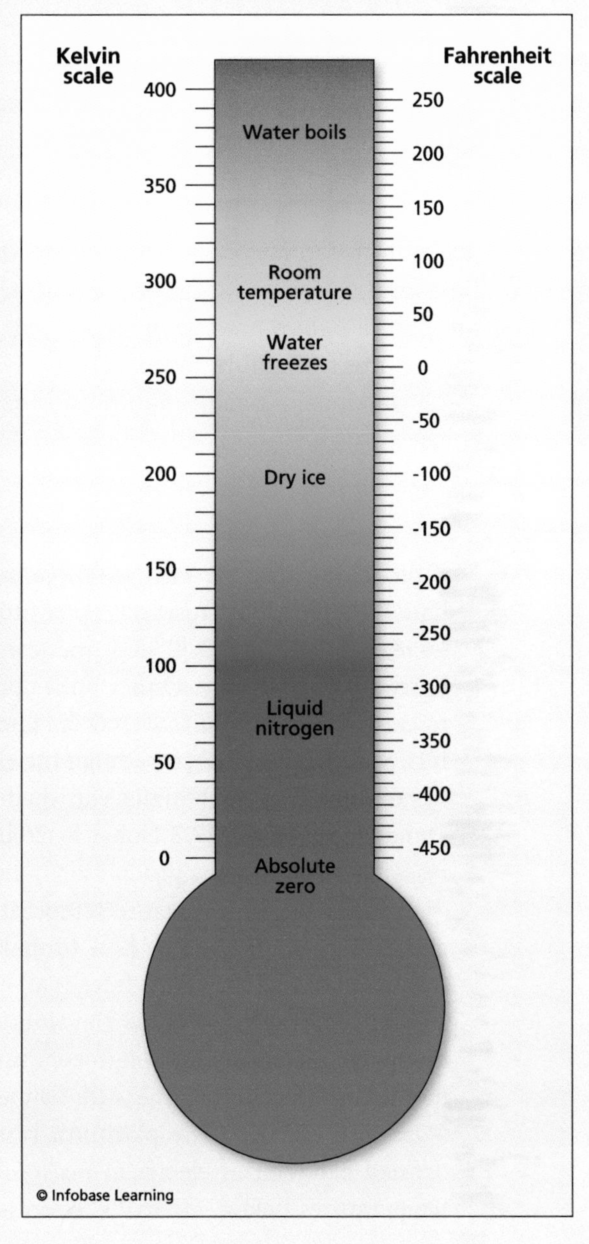


**This figure compares the absolute Kelvin temperature scale with the relative Fahrenheit temperature scale. Several important reference points (assuming an ambient pressure of one atmosphere) are also indicated, including absolute zero.** ***(Modeled after NASA/JPL)***

*(continued)*

microkelvin (μK) ($10^{-6}$ K), nanokelvin (nK) ($10^{-9}$), and even picokelvin (pK) ($10^{-12}$) environments. For example, a team of NASA-funded scientists reported in 2003 that they created a Bose–Einstein condensate (BEC) at an incredibly frigid temperature of only 450 pK. The Bose–Einstein condensate (BEC) is state of matter in which extremely cold atoms attain the same quantum state and behave like a large superatom. Satyendra Nath Bose (1894–1974) was an Indian mathematical physicist who collaborated with Albert Einstein and made important contributions to quantum mechanics.

As an integral part of the notion of absolute zero, scientists identify the absolute temperature of an object. Honoring Lord Kelvin's contributions, they express absolute temperature in kelvins (K), the fundamental SI unit of temperature. Around the world, engineers also use kelvins to express absolute temperatures in radiation heat transfer computations, but American engineers occasionally express absolute temperatures in rankines (R). One kelvin corresponds to 1.8 rankines. Absolute temperatures are very important in the field of cryogenics.

approaching absolute zero and became the first scientist to successfully liquefy helium. His research apparatus achieved a temperature of just −458.05°F (−272.25°C [0.9K]), the lowest temperature ever achieved on Earth up to that time. Onnes built upon his successful low-temperature research efforts and discovered the phenomenon of superconductivity in 1911. Specifically, he observed that the electrical resistance in a test sample of mercury (Hg) essentially vanished at −452.11°F (−268.95°C [4.2 K]). Onnes received the 1913 Nobel Prize in physics for his low-temperature research, including the production of liquid helium.

As they began to explore temperatures in the cryogenic range, scientists were challenged by how to make accurate temperature measurements. Since mercury (Hg) freezes at −37.892°F (−38.83°C [234.32 K]), the substance became useless as the working fluid in low-temperature thermometers. Scientists and engineers turned to platinum resistance thermometers to accurately measure temperatures down to about −423.67°F, (−253.15°C [20 K]). The platinum resistance thermometer has a well-defined behavior of electrical resistance versus temperature. To record temperatures below −423.67°F, scientists used electrical resistance thermometers made of certain semiconductor materials such as doped germanium. With this approach, they were able to measure temperatures down to −457.87°F (−272.15°C [1K]) and below. Low-temperature thermometers based on such semiconductor materials require that a physical variable such as electrical resistivity change in a well-known theoretical way with respect to temperature.

The production of cryogenic temperature regimes starts with the compression and expansion of gases. For example, in a typical air liquefaction process, engineers compress atmospheric air, causing it to heat. They then allow the compressed air to cool in a heat exchanger before rapidly expanding the gas back to atmospheric pressure. This rapid expansion in a closed container allows the air to cool even more, transforming a portion of the air into a liquid. Liquefied air is pale blue and contains mostly nitrogen and oxygen. At one atmosphere pressure, oxygen has a boiling point of –297.39°F (–183.0°C [90.2 K]) and nitrogen a boiling point of –321°F (–196.2°C [77 K]). The cooled gaseous portion of the expanded air returns to the mechanical compression and expansion process, while the liquid air is distilled to yield liquid oxygen, liquid nitrogen, and other atmospheric components, such as liquid argon. Scientists use gaseous helium to produce even lower cryogenic temperature environments, but several stages of mechanical compression-expansion are necessary.

There are two other physical processes that scientists and engineers use to create temperatures near absolute zero. These are the Joule-Thomson (Kelvin) effect and adiabatic demagnetization. While collaborating in Great Britain in 1852, James Prescott Joule and Lord Kelvin (William Thomson) discovered that when a real (nonperfect) gas expands through a throttling device such as a nozzle or a porous plug from a high-pressure region to a low-pressure region, the temperature of the gas drops. Using this phenomenon, engineers have developed practical refrigerators. The Joule-Thomson effect also serves as a key step in the liquefaction of gases such as helium. When Onnes first liquefied helium in 1908, he initially chilled the helium gas by contacting it with liquid oxygen, then liquid nitrogen, and finally liquid hydrogen. Onnes then expanded the very cold helium gas through a Joule-Thomson nozzle and produced a mixture of frigid helium gas and liquid droplets.

A great deal of interesting thermodynamic theory is needed to understand why real gases experience Joule-Thomson effect cooling upon

*(continues on page 166)*

## LIQUID PROPELLANT ROCKETS

The American physicist Robert Hutchings Goddard (1882–1945) invented liquid propellant in 1926. So important were his contributions to the field of modern rocketry that it is appropriate to state "Every liquid propellant rocket is essentially a Goddard rocket."

The accompanying illustration describes the major components of a liquid bipropellant chemical rocket. The liquid propellant rocket's propulsion system contains propellant tanks, the combustion chamber, the nozzle assembly, and turbopumps. Aerospace engineers design the propellant tanks as load-bearing structures capable of containing the rocket's liquid propellants. There is a separate tank for the fuel and for the oxidizer. The combustion chamber is the region into which the liquid propellants are pumped, vaporized, and reacted (combusted) to create hot exhaust gases, which then expand through the nozzle and generate thrust. Turbopumps are precision fluid flow machinery that deliver propellants from the tanks into the combustion chamber at high pressure and sufficient mass flow rate (symbol: $\dot{m}$). The thrust equation (also shown in the illustration) is the fundamental equation of rocket performance.

Newton's third law of motion, the action-reaction principle, is the basis of the operation of all rockets. As the rocket engine expels mass at high velocity, the resulting reaction thrust (or force F) drives the rocket vehicle in the opposite direction. For reaction engines, such as rockets, that generate thrust *(T)* by expelling a stream of internally carried mass at a mass flow rate $\dot{m}$), scientists write this basic equation as $T = \dot{m}\ V_e\ (p_e - p_0)\ A_e$, where $V_e$ is the exhaust velocity of the ejected mass, $p_e$ is the pressure of the hot combustion gases as they exit the nozzle, $p_0$ is the ambient pressure (note that rockets work best in the vacuum of space, where $p_0 = 0$), and $A_e$ is the nozzle's exit area.

Aerospace engineers know that cryogenic propellant combinations such as liquid hydrogen for the fuel and liquid oxygen (LOX) for the oxidizer result in very high performance chemical rockets. These engineers also recognize that they must design cryogenic propellant tanks with extensive insulation to keep the propellants in liquefied form prior to launch. Unfortunately, with even the best available insulation, cryogenic propellants are difficult to store for long periods of time and eventually will boil off (that is, vaporize). At normal sea level pressure, oxygen remains in the liquid state when its temperature is kept

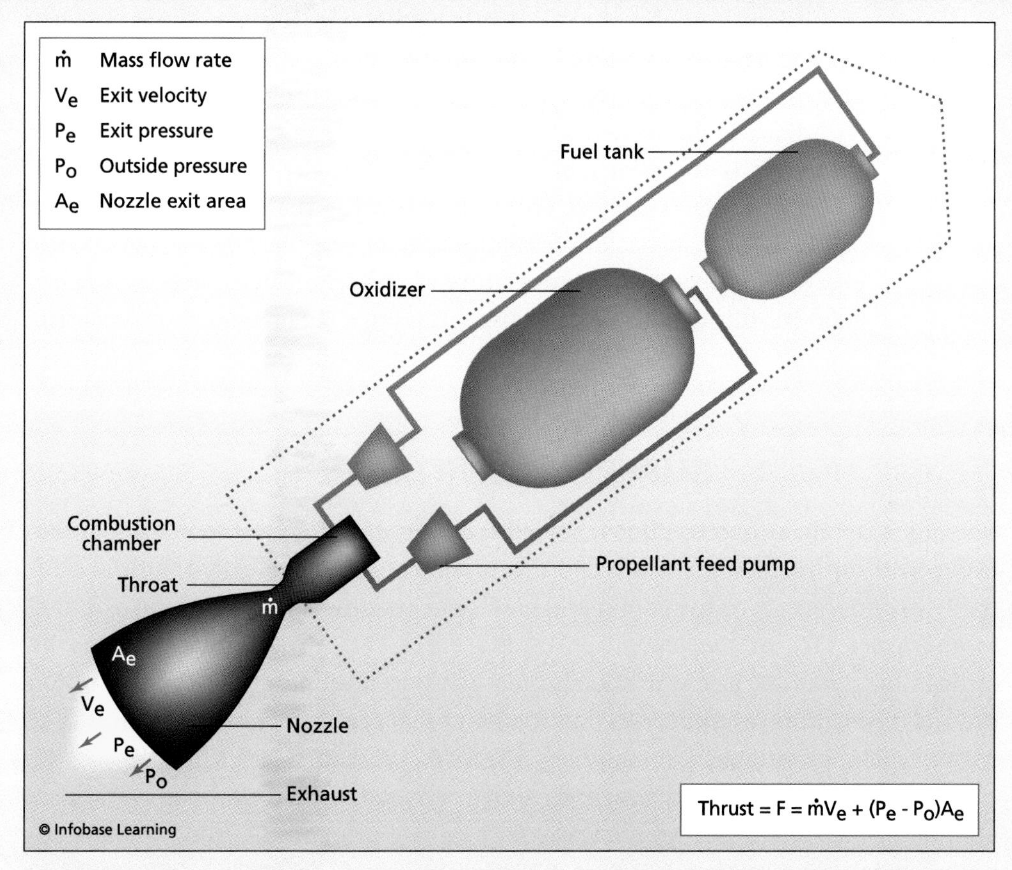


**This diagram depicts the major components of a pumped liquid bipropellant rocket. The accompanying equation is rocketry's famous thrust equation.** ***(Based on NASA artwork)***

at –297°F (–183°C [90.2 K]) or below, while hydrogen remains liquid at –423°F (–253°C [20.2 K]) or below.

Rocket scientists do not want vapor pressures to build up in the cryogenic liquid propellant tanks, nor do they want to store the high-performance propellants as high-pressure gases. Either undesirable condition requires the use of more structural material for the propellant tanks and thereby reduces the overall payload a given rocket system can deliver. Venting gaseous hydrogen or gaseous oxygen from a propellant tank at the launch site also represents a major safety hazard, so engineers approach the design of cryogenic propellant tanks with a very simple rule: Keep it cool, very cool.

*(continued from page 163)*
expansion through a throttle, while perfect (ideal) gases do not. The key point is that refrigeration devices that employ the Joule-Thomson effect allow scientists to achieve temperatures approaching absolute zero.

Adiabatic demagnetization refrigerators (ADRs) are routinely used by scientists to reach temperatures below 1 K, that is, below–457.9°F (–272.2°C). These devices are based upon the thermodynamic process called the magnetocaloric effect. Scientists describe the magnetocaloric effect as the decrease in temperature when certain substances experience adiabatic demagnetization. To use this interesting phenomenon, scientists

## SUPERCONDUCTIVITY

Scientists define superconductivity as the ability of a material to conduct electricity without resistance. They call the temperature at which a material suddenly experiences zero (or negligible) electrical resistivity that material's critical temperature ($T_C$), or the transition temperature ($T_T$). When cooled below its critical temperature, a material becomes a superconductor. In 1911, the Dutch physicist Heike Onnes observed that the electrical resistance in a test sample of mercury (Hg) essentially vanished at –452.11°F (–268.95°C [4.2 K]).

Other scientists soon discovered that many other metals become superconductors at temperatures near absolute zero. For example, the critical temperature of aluminum is –457.5°F (–272°C [1.18K]), tin is –453°F (–269.4°C [3.72 K]), lead is –443°F (–263.9°C [9.26 K]), and niobium is –447°F (–266°C [7.04 K]). Commercially available superconducting coils, often composed of a niobium-titanium alloy (NbTi), create the very powerful magnets that scientists use in particle accelerators and that medical engineers incorporate in magnetic resonance imaging (MRI) systems. The critical temperature of NbTi alloy is –441.7°F (–263.2°C [10 K]). Engineers plan to use superconductor materials to achieve the magnetic levitation of high-speed trains and in the development of more efficient transmission lines for electric power distribution. One difficulty is efficiently cooling metallic superconductor materials that have critical temperatures near absolute zero.

In the 1980s, the German physicist Johannes Georg Bednorz (1950– ) and the Swiss physicist Karl Alexander Müller (1927– ) started a system-first use liquid helium to cool a paramagnetic salt down to a few kelvins. Once the paramagnetic salt has been properly chilled, scientists apply a strong magnetic field, then they thermally isolate the paramagnetic salt and reduce the applied magnetic field to zero. Due to the magnetocaloric effect, the temperature of the paramagnetic salt drops to one kelvin (1 K) or less. Physicists regard a paramagnetic (weakly magnetic) substance as one that has an assembly of magnetic dipoles that have random orientations. Ferric ammonium sulfate ($FeNH_4(SO_4)_2 \cdot 12H_2O$) is a paramagnetic salt that NASA engineers have used to construct a cyclic magnetic cooler, or ADR.

atic investigation of the electrical properties of certain ceramic materials. By 1986, they had succeeded in demonstrating that lanthanum barium copper oxide (LaBaCuO, or LBCO) exhibited superconductivity at a critical temperature of −396.7°F (−238.2°C [35 K]). This was an important milestone in the development of high-temperature superconductor materials. For their discovery of superconductivity in ceramic materials, Bednorz and Müller shared the 1987 Nobel Prize in physics. Other researchers have since identified superconductor materials that have critical temperatures near or even above the boiling point of liquid nitrogen (−321°F [−196.2°C {77 K}]). For example, one high-temperature copper oxide superconductor ($HgBa_2Ca_2Cu_3Ox$) has a critical temperature of −216.7°F (−138.2°C [135 K]). Ceramic materials that function at or above liquid nitrogen temperatures significantly expand the beneficial applications of superconductivity. Compared to liquid helium, liquid nitrogen is a more easily handled and much less expensive coolant.

**A magnet is levitated by high-temperature (copper oxide) superconducting materials that are cooled in liquid nitrogen.** ***(DOE)***

## APPLICATIONS OF CRYOGENIC LIQUIDS

Cryogenic liquids enjoy many applications. As discussed in this section, there are numerous military, industrial, and medical uses for liquid oxygen and liquid nitrogen. Liquid hydrogen has a well-proven role in space exploration as a chemical rocket fuel and an equally significant role in future global energy infrastructure. Liquid helium plays a very special role in scientific research, space exploration, and superconductivity.

### Liquid Oxygen (LOX)

Liquid oxygen has a boiling point of –297°F (–183°C [90.2 K]) at a pressure of one atmosphere. Because oxygen is so reactive, the pale blue liquid must be stored in clean systems that are constructed of high-ignition temperature, nonreactive materials. Oxygen is often stored as a liquid, although it is used primarily as a gas. For example, in the U.S. Air Force, flight support personnel upload liquid oxygen (LOX) on certain military

A U.S. Air Force cryogenics technician services a liquid oxygen (LOX) cart at Aviano Air Base, Italy. *(USAF)*

aircraft. The liquid oxygen is then converted into gaseous form to provide pure oxygen to pilots during high-altitude flight.

Personnel who handle liquid oxygen or any other cryogenic liquid must do so with great caution because they can expose themselves to very cold temperatures and suffer severe freeze burns. They should use personal protection equipment, including a full face shield over safety glasses, thermally insulated gloves, safety shoes, and proper laboratory clothing. The gloves should be loose-fitting so they can be quickly removed in the event that the cryogenic fluid spills or splashes on them.

Containers for cryogenic liquids, including liquid oxygen, should have reliable pressure relief valves. Adequate ventilation is also essential. Without adequate ventilation, the excessive accumulation of gaseous oxygen due to boil-off and venting from a liquid oxygen storage system poses a serious hazard. If the vented oxygen accumulates and enriches the surrounding atmosphere, there is a significant threat of a violent combustion reaction. Substances that readily ignite in air become more susceptible to ignition in the presence of liquid oxygen or an oxygen-enriched atmosphere. Material surfaces splashed with liquid oxygen, including clothing, become more sensitive to ignition and will combust vigorously. Finally, the use of any open flame, including smoking, must be strictly prohibited in areas where liquid oxygen is stored, handled, or used.

## Liquid Nitrogen ($LN_2$)

As an inexpensive, inert cryogen, liquid nitrogen ($LN_2$) enjoys many applications in research, industry, and medicine. At a pressure of one atmosphere, liquid nitrogen boils at –321°F (–196.2°C [77 K]). Like liquid oxygen, liquid nitrogen is harvested from the atmosphere by modern air liquefaction processes. Liquid nitrogen is colorless, odorless, inert, noncorrosive, and nonflammable. However, the cryogen is very cold, so safety procedures and personal protection equipment are essential if severe skin burns and frostbite are to be avoided. Because the boiling point of liquid oxygen is above that of liquid nitrogen, liquid nitrogen should not be stored for long periods in an open container. Over time, oxygen can condense from the surrounding air into the liquid nitrogen. If the room air over the liquid nitrogen circulates, as occurs in many laboratory, academic, and industrial facilities, condensed liquid oxygen can build up to violently combustible levels. Organic materials, including clothing, that are ordinarily nonflammable may suddenly ignite in a room environment enriched by condensed liquid oxygen.

Cryogenic containers for liquid nitrogen must be equipped with a pressure relief valve to control internal pressure. Under normal conditions, these cryogenic containers will vent small quantities of gaseous nitrogen. Areas where liquid nitrogen is stored must be well ventilated, or else suffocation can occur. Although nitrogen is nontoxic and inert, the gas can act as a simple asphyxiant by displacing oxygen in the air to levels below those required to support life. The normal oxygen concentration in air is 21 percent by volume. Inhalation of nitrogen gas in excessive amounts can cause dizziness, nausea, vomiting, loss of consciousness, and death. At low oxygen concentrations in a working environment, unconsciousness and death may occur in seconds and without warning. If venting nitrogen gas from a cryogenic storage device can or has caused the concentration of oxygen in a confined workplace to fall below 19.5 percent by volume, workers must each use a self-contained breathing apparatus.

Aerospace engineers use liquid nitrogen to simulate the frigid temperature conditions that spacecraft can encounter during a mission in outer space. The Mark-1 Test Facility at the Arnold Engineering Development Center of the U.S. Air Force is an example. This state-of-the-art space environment simulation facility accommodates the testing of full-scale space systems. With a working interior diameter of 36 feet (11 m) and a height of 70 feet (21.3 m), large aerospace systems can be installed for testing, and smaller objects can use the height to experience micro-gravity conditions for two seconds. The huge test chamber is lined with a liquid nitrogen shroud that provides thermal conditions down to –321°F (–196.2°C [77 K]).

In cryosurgery, physicians use liquid nitrogen to selectively remove unnecessary or diseased tissue. The physician employs a special precision instrument, through which liquid nitrogen circulates. Upon contact with tip of the intensely cold probe (maintained at –321°F [–196.2°C {77 K}]) the target tissue is frozen and destroyed. Doctors use this generally bloodless form of surgery to treat warts and destroy certain types of tumors. Neurosurgeons use liquid nitrogen–cooled probes to treat certain brain disorders.

The practice of cryopreservation involves the use of very low temperatures to preserve indefinitely cells, tissues, blood products, skin, embryos, sperm, and similar biological materials. Scientists generally draw on liquid nitrogen to create and maintain the required low-temperature environment. Rigorous protocols are necessary for freezing and later thawing viable biological materials. These protocols vary with the biological material being preserved.

An officer of the U.S. Air Force visits inside the huge Mark-1 Space Chamber at the Arnold Engineering Development Center in Tennessee. When operated, this space environment simulation facility can test fully assembled spacecraft. The chamber's liquid nitrogen shroud exposes aerospace systems to thermal conditions down to –321°F (–196.2°C [77 K]). *(USAF)*

A plant physiologist lowers a container of seeds into the vat of liquid nitrogen that will preserve them. The longevity of seeds stored through cryopreservation is projected to be thousands of years, but agricultural scientists periodically remove samples to assess their viability and vigor. *(USDA)*

Plant physiologists use cryopreservation to store seeds of plants important to human survival. They also preserve seeds from rare and endangered plants. Similarly, veterinary scientists use cryopreservation to store the embryos and sperm of endangered animals or prized livestock. Finally, human fertility clinics help men and women who wish to use cryopreservation techniques to store their sperm or oocytes (eggs). When in vitro fertilization procedures result in more fertilized embryos than are immediately required, the physician may recommend cryopreservation to store the surplus embryos for later use.

Medical specialists use controlled-rate freezing techniques to slowly cool embryos placed in a special cryoprotectant fluid down from body temperature to –321°F (–196.2°C [77 K]). Once at this low temperature, the embryos are stored in special containers within a bath of liquid nitrogen. In 1983, the medical profession reported the first successful pregnancy from a frozen/thawed human embryo. By 2000, some 16,000 cases of assisted reproductive technology (ART) in the United States involved the use of frozen/thawed human embryos. At present, medical studies of the human offspring who arise from frozen/thawed embryos have not revealed any significant increase in birth defects or physical abnormalities compared to the pregnancy outcomes in the rest of the population. The medical profession has developed ethical guidelines concerning human embryo cryopreservation.

## Liquid Hydrogen ($LH_2$)

Liquid hydrogen has a boiling point of –423.2°F (–252.9°C [20.3 K]). Although hydrogen is the most abundant element in the universe, it is not found alone as an element on Earth. Liquid hydrogen is often used along with liquid oxygen as the propellants in modern rocket vehicles. Liquid hydrogen also plays a role in the study and application of high-temperature superconductors. Energy engineers regard liquid hydrogen as an important energy carrier in the future.

An energy carrier is a substance or system that moves energy in a usable form from one place to another. Today, electricity is the most well known energy carrier. People use electricity to transport the energy available in coal, uranium, and falling water from generation facilities to points of application such as industrial sites and homes. An energy carrier makes the application of the energy content of the primary source more convenient. Liquid hydrogen can store energy until it is needed and can also move energy to a variety of places where it is needed.

Since hydrogen is not found as a free element on Earth, engineers have developed processes to separate it from other elements. At present, the two most common methods for producing hydrogen are steam reforming and electrolysis. The least expensive method is steam reforming, which accounts for about 95 percent of the hydrogen produced in the United States. In this process, hydrogen atoms are separated from the carbon atoms in methane ($CH_4$) molecules. Typically, chemical engineers employ an external source of hot gas to heat tubes in which a catalytic reaction takes place, converting steam (very hot $H_2O$) and lighter hydrocarbons, such as natural gas (methane [$CH_4$]), into hydrogen and carbon monoxide (CO).

In the other production process, electrolysis, hydrogen atoms are harvested after water molecules ($H_2O$) are split. Electrolysis is currently very costly, but engineers are examining future approaches to hydrogen production at large central plants or small local plants. Once liquid hydrogen becomes less expensive and more available, it can be used in transportation systems as an environmentally friendly, combustible fuel, since the combustion by-product is water. Hydrogen can also be used in fuel cells to generate power for vehicles, homes, and industrial sites.

As an extremely cold cryogen, proper equipment is needed to safely store and handle liquid hydrogen. Hydrogen gas is flammable, so care must also be exercised to avoid situations where gas leaks represent an explosive hazard. The flammable range of hydrogen in room temperature air at one atmosphere pressure is 4 to 75 percent by volume. The flame temperature of hydrogen in air is 3,713°F (2,045°C).

The American aerospace program has successfully handled liquid hydrogen for decades. Operational and safety procedures that involve the space program's use of liquid hydrogen represent an important technical legacy. These experiences can assist in the growth of a global hydrogen-based energy economy this century.

## Liquid Helium

Liquid helium is the most important scientific cryogen and plays a special role in research, space exploration, and superconductivity. Early in the 20th century, the Dutch physicist Heike Onnes developed experimental techniques to investigate temperatures approaching absolute zero. In 1908, he became the first scientist to successfully liquefy helium. His research apparatus reached a temperature of just −458.05°F (−272.25°C [0.9K], the lowest temperature ever achieved on Earth up to that time. Three years later, Onnes observed that at the frigid temperature of −452.11°F (−268.95°C [4.2 K]), the electrical resistance in a test sample of mercury (Hg) essentially vanished. He had discovered the important phenomenon of superconductivity.

Although helium is the second-most abundant element in the universe, its existence was totally unknown prior to the late 19th century. In 1868, the British physicist Sir Joseph Norman Lockyer (1836–1920) collaborated with the French astronomer Pierre Jules César Janssen (1824–1907) and discovered the element helium through spectroscopic studies of solar prominences. Specifically, while observing a total solar eclipse, Janssen noticed a peculiar yellow line in the solar spectrum. Lockyer soon recognized that this spectral line at a wavelength of 587.49 nm could not be produced by any element then known, so he postulated the existence of a new element, which he named helium—after *helios* (ἥλιος), the sun god in ancient Greek mythology.

However, it was not until 1895 that the Scottish chemist Sir William Ramsay (1852–1916) finally detected helium on Earth. He made this discovery while carefully examining the uranium mineral cleveite. The gaseous helium Ramsay detected came from the radioactive decay of uranium and was trapped within the uranium-bearing mineral. Today, helium gas is commercially recovered from natural gas deposits found primarily in Texas, Oklahoma, and Kansas. Helium makes up about 0.0005 percent of Earth's atmosphere. This trace amount is constantly being lost to space. However, the alpha decay of radioactive elements in Earth's crust constantly releases helium, some of which seeps up through cracks in the crust and replenishes the trace quantity found in the atmosphere.

As normally encountered here on Earth, helium is a colorless, odorless, inert, and nontoxic gas that resides at the top of the noble gas family in the periodic table. (See appendix.) Helium has the lowest boiling and melting points of all the known elements, so the substance exists as a liquid or solid only under extreme conditions of low temperature and high pressure. Unlike any other known element, helium remains a liquid down to absolute zero under normal conditions of pressure.

To appreciate the unusual properties of liquid helium, it is first necessary to recognize that helium has two natural, stable isotopes: helium-4 and helium-3. With a natural abundance as found on Earth of 99.999863 percent, helium-4 is by far the more prevalent form of helium. The nucleus of a helium-4 atom consists of two protons and two neutrons. The nucleus of the helium-3 atom consists of two protons and only one neutron. With a natural abundance of just 0.000137 percent, helium-3 is an extremely rare natural isotope of helium. Quantum mechanical differences in the nuclei of these two stable isotopes result in some very interesting behaviors as cryogenic fluids. This section primarily addresses the cryogenic behavior of helium-4, but certain properties of helium-3 as a cryogenic fluid will also be mentioned.

Helium-4 has a boiling point of –452.1°F (–268.95°C [4.2 K]) at one atmosphere pressure, while helium-3 has a boiling point of –453.9°F (–269.95 °C [3.2 K]) at one atmosphere pressure. Unlike any other elemental substance, helium remains a liquid down to absolute zero at normal pressures. The only way scientists are able to transform liquid helium into the solid state is to increase the pressure on the fluid while keeping it at an extremely low temperature. The Dutch physicist Willem Hendrik Keesom (1876–1956) developed a technique to solidify helium in 1926. Scientists now understand that helium-4 transforms from a liquid to a solid at a temperature of about –457.9°F (–272.15°C [1 K]) and a pressure of 25 atmospheres, while helium-3 experiences this transition at about 0.3 K and a pressure of 29 atmospheres.

While experimenting with liquid helium-4 at extremely low temperatures in the 1930s, scientists discovered an amazing phenomenon called superfluidity. They observed that between 4.22 K and 2.177 K and at one atmosphere pressure, the isotope helium-4 was a very cold, colorless liquid. (Recall that 4.22 K is the boiling point of helium-4 at one atmosphere pressure.) Scientists called this form of liquid helium-4 helium I, often abbreviated He-4I. Like other cryogenic liquids, helium I boils when heated and contracts when the temperature is lowered. This behavior continues until helium I reaches 2.177 K.

At 2.177K, helium I experiences a remarkable physical discontinuity. The thermal conductivity of the cryogenic liquid increases dramatically, and its viscosity becomes zero. Scientists refer to the temperature associated with this remarkable transition as the lambda point; they call liquid helium cooled below the lambda point a superfluid. Also termed *helium II*

## SUPERFLUIDS

Scientists define a superfluid as a zero viscosity liquid that flows with no inner friction. Physicists further note that a liquid becomes a superfluid when all its atoms or molecules have been cooled (or condensed) to the point at which they all occupy the same quantum state. Toward the end of the 1930s, the Russian physicist Pyotr Kapista (1894–1984) began a series of experiments in low-temperature physics that involved the study of liquid helium. His work led to the discovery in 1937 of superfluidity involving the form of liquid helium-4 called helium II. He received a portion of the 1978 Nobel Prize in physics for his pioneering low-temperature research. Kapista's work also inspired another Russian physicist, Lev Davidovich Landau (1908–68). Landau performed important theoretical studies concerning the unusual behavior of liquid helium at near absolute zero. Landau received the 1962 Nobel Prize in physics for his pioneering theories of condensed matter, especially as related to the superfluid behavior of liquid helium-4.

Such research efforts clearly indicated that classical physics was inadequate to explain the phenomenon of superfluidity. Superfluids exhibit a variety of unusual, nonclassical behaviors, including flow through very small cracks and creep flow up and out of containers. Very cold liquids that exhibit superfluidity are often termed *quantum fluids* because physicists must use advanced forms of quantum mechanics to explain their behavior. Today, scientists typically refer to a superfluid as a quantum fluid with perfect microscopic (that is, atomic-level) order, or zero entropy.

In 1996, three American physicists—David M. Lee (1931– ), Douglas D. Osheroff (1945– ), and Robert C. Richardson (1937– )—shared the Nobel Prize in physics for their discovery that liquid helium-3 exhibited superfluidity at a temperature of about 0.0025 K. Research suggests that as a quantum fluid, liquid helium-3 possesses a significantly more complicated nuclear structure than the quantum fluid liquid helium-4. For example, liquid helium-3 is anisotropic, meaning the quantum fluid displays different physical properties in different spatial directions. The British-American physicist Sir Anthony James Leggett (1938– ) received the 2003 Nobel Prize in physics for his theoretical work concerning the superfluidity of helium-3.

(He-4II), this form of liquid helium is unlike any other known substance. Below the lambda point, helium II does not boil and exhibits superfluid behavior. This means helium II can flow through microscopic channels and very tiny capillaries and creep out of supposedly leak-tight containers. Cryogenic engineers discovered that they had to design very special containers to store helium II. Otherwise, the superfluid would "creep" along the container's surfaces and "sneak" through valves and microscopic cracks until it reached a warmer region, where it would evaporate.

Liquid helium plays an essential role in modern science, including the operation of large particle accelerators that rely on superconductors and the cooling of special instruments on spacecraft. For example, in 1989, NASA successfully launched the *Cosmic Background Explorer (COBE)* into orbit around Earth. Instruments on this scientific spacecraft provided the scientific community with its first definitive measurement of the cosmic microwave background, thereby confirming the big bang theory of the origin of the universe. The *COBE* spacecraft carried a cryostat filled with 174 gallons (660 L) of liquid helium. The liquid helium provided a stable −457.15°F (−271.75°C [1.4 K]) environment for two of the spacecraft's instruments: the far infrared absolute spectrometer (FIRAS) and the diffuse infrared background experiment (DIRBE). On September 21, 1990, after 306 days of successful cryogenic operations, the last of the superfluid helium contained in the dewar was consumed, and the initial phase of this important space mission came to an end. The American scientists John C. Mather (1946–    ) and George F. Smoot (1945–    ) received the 2006 Nobel Prize in physics for their discovery of the blackbody form and anisotropy of the cosmic microwave background.

# CHAPTER 11

# Conclusion

**Liquids occupy a special niche in the world (hierarchy) of matter. More** "mobile and playful" than rigid solids, liquids are not quite as "flighty or nomadic" as gases. One liquid above all, water, dominates Earth's ability to host an incredible treasury of living creatures. Covering approximately 71 percent of the planet's surface and containing 97 percent of the planet's water, Earth's majestic oceans contain nearly 50 percent of all species found on Earth and provide about 5 percent of the total protein in the human diet.

Freshwater is essential for life on Earth's surface. The earliest civilizations became known as hydrocivilzations because they emerged along rivers. Ancient peoples manipulated the flow of water, produced wine and beer, and used mercury to extract gold and silver from ore. Nevertheless, the scientific understanding of what actually happened when water vaporizes and becomes steam, how ocean currents govern the planet's climate, and how blood flows through the human body awaited scientific interpretations of liquid behavior.

The ability of human beings to relate the microscopic (atomic level) behavior of liquid matter to readily observable macroscopic properties (such as density, pressure, temperature, and viscosity) transformed science and engineering. Archimedes and other great engineers of antiquity used liquids effectively without possessing a detailed knowledge of their microscopic composition. The real breakthroughs in accurately describing the different properties of liquid matter started during the Scientific Revolution.

During the Scientific Revolution, pioneering scientists such as Galileo Galilei, Sir Isaac Newton, Blaise Pascal, and Daniel Bernoulli began describing and predicting fluid behavior. Their technical efforts yielded important fluid science relationships for both liquids and gases. They employed interesting experiments and mathematical relationships to help unlock nature's secrets. Their pioneering activities formed the broad field of fluid science.

Fluid science principles and concepts govern many fields, including astronomy, biology, medical sciences, chemistry, Earth science, meteorology, geology, oceanography, and physics. Fluid science forms an integral part of many branches of engineering, including aeronautics and astronautics, biological engineering, chemical engineering (including petroleum extraction and refining), civil engineering (including dam construction and calculating wind loads on tall buildings), environmental and sanitary engineering, industrial engineering (including food and beverage processing), mechanical engineering, nuclear engineering, and ocean engineering (including shipbuilding).

Modern water, sanitation, and hygiene systems are one of humankind's great engineering achievements, yet there is still much room for improvement on a global basis. Safe drinking water, sanitation, and hygiene represent the three most important conditions for keeping communities healthy. According to the World Health Organization, unsafe drinking water coupled with a lack of basic sanitation kill at least 1.6 million children under the age of five each year. Furthermore, about 2.6 billion people around the world do not use a toilet—they defecate in open fields or other unsanitary places.

Petroleum and mercury are ubiquitous and essential liquids in the modern world. Petroleum, also referred to as black gold, is the energetic lifeblood of modern civilization, while mercury has widespread uses in scientific instruments, personal electronics, streetlights, fluorescent lamps, dry cell batteries, and even dental fillings. When properly used, controlled, and disposed of, both petroleum and mercury are marvelous technical servants of the human race. However, when haphazardly released into the environment by accident or intention, they can cause serious harm.

Cryogenic liquids are essential in science and technology. Superconducting magnets support the operation of powerful particle accelerators that probe the mysteries of matter and form an integral part of magnetic resonance imaging tools used in modern medicine. Liquid helium plays an essential role in modern science, including the operation of large particle accelerators that rely on superconductors and the cooling of special instruments on spacecraft. For example, in 1989, NASA successfully launched the *Cosmic Background Explorer (COBE)* into orbit around Earth. Instruments on this scientific spacecraft provided the scientific community its first definitive measurement of the cosmic microwave background, thereby confirming the big bang theory of the origin of the universe.

The American aerospace program has successfully handled liquid hydrogen for decades. Operational and safety procedures that involve the space program's use of liquid hydrogen constitute an important technical legacy. These experiences can assist the growth of a global hydrogen-based energy economy in this century.

# Appendix

Scientists correlate the properties of the elements portrayed in the periodic table with their electron configurations. Since, in a neutral atom, the number of electrons equals the number of protons, they arrange the elements in order of their increasing atomic number (Z). The modern periodic table has seven horizontal rows (called periods) and 18 vertical columns (called groups). The properties of the elements in a particular row vary across it, providing the concept of periodicity.

There are several versions of the periodic table used in modern science. The International Union of Pure and Applied Chemistry (IUPAC)

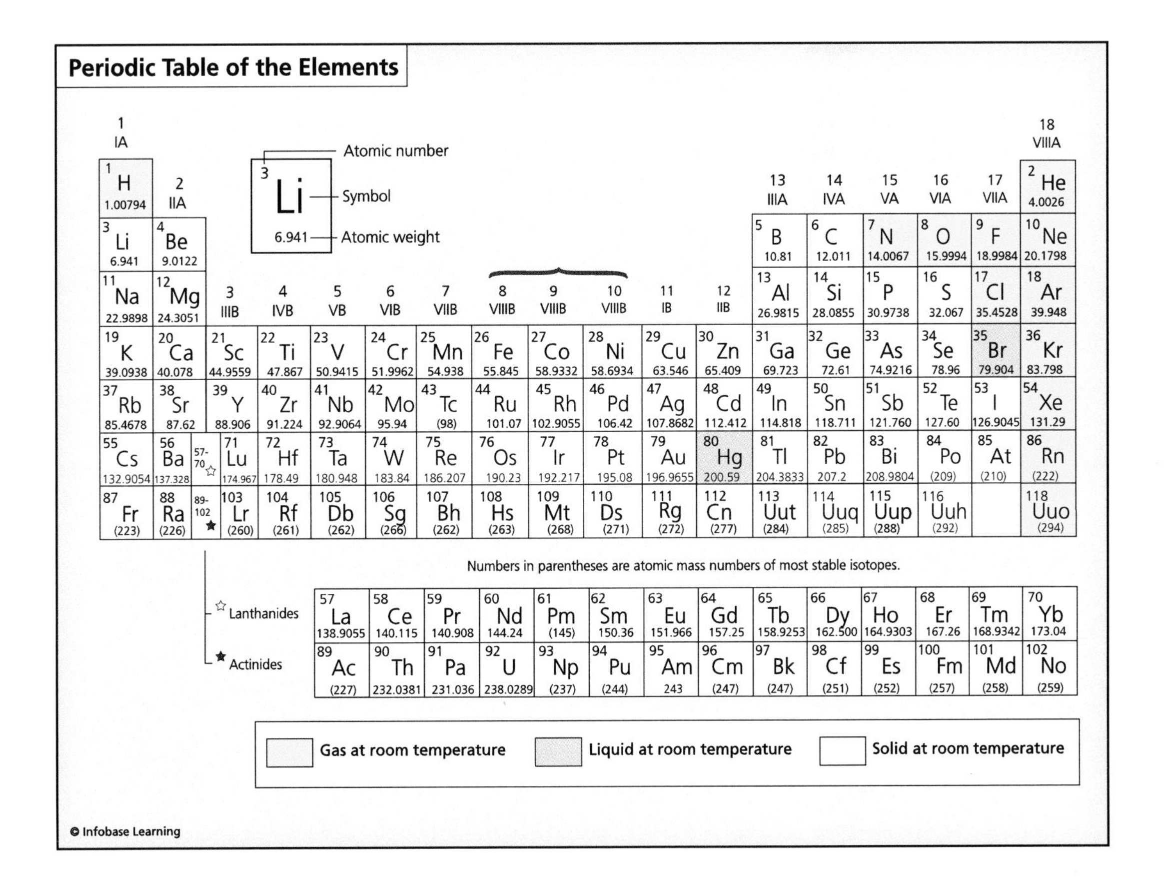

recommends labeling the vertical columns from 1 to 18, starting with hydrogen (H) as the top of group 1 and ending with helium (He) as the top of group 18. The IUPAC further recommends labeling the periods (rows) from 1 to 7. Hydrogen (H) and helium (He) are the only two elements found in period (row) 1. Period 7 starts with francium (Fr) and includes the actinide series as well as the transactinides (very short-lived, human-made, super-heavy elements).

The row (or period) in which an element appears in the periodic table tells scientists how many electron shells an atom of that particular element possesses. The column (or group) lets scientists know how many electrons to expect in an element's outermost electron shell. Scientists call an electron residing in an atom's outermost shell a valence electron. Chemists have learned that it is these valence electrons that determine the chemistry of a particular element. The periodic table is structured such that all the elements in the same column (group) have the same number of valence electrons. The elements that appear in a particular column (group) display similar chemistry.

## ELEMENTS LISTED BY ATOMIC NUMBER

| | | | | | |
|---|---|---|---|---|---|
| 1 | H | Hydrogen | 14 | Si | Silicon |
| 2 | He | Helium | 15 | P | Phosphorus |
| 3 | Li | Lithium | 16 | S | Sulfur |
| 4 | Be | Beryllium | 17 | Cl | Chlorine |
| 5 | B | Boron | 18 | Ar | Argon |
| 6 | C | Carbon | 19 | K | Potassium |
| 7 | N | Nitrogen | 20 | Ca | Calcium |
| 8 | O | Oxygen | 21 | Sc | Scandium |
| 9 | F | Fluorine | 22 | Ti | Titanium |
| 10 | Ne | Neon | 23 | V | Vanadium |
| 11 | Na | Sodium | 24 | Cr | Chromium |
| 12 | Mg | Magnesium | 25 | Mn | Manganese |
| 13 | Al | Aluminum | 26 | Fe | Iron |

| | | |
|---|---|---|
| 27 | Co | Cobalt |
| 28 | Ni | Nickel |
| 29 | Cu | Copper |
| 30 | Zn | Zinc |
| 31 | Ga | Gallium |
| 32 | Ge | Germanium |
| 33 | As | Arsenic |
| 34 | Se | Selenium |
| 35 | Br | Bromine |
| 36 | Kr | Krypton |
| 37 | Rb | Rubidium |
| 38 | Sr | Strontium |
| 39 | Y | Yttrium |
| 40 | Zr | Zirconium |
| 41 | Nb | Niobium |
| 42 | Mo | Molybdenum |
| 43 | Tc | Technetium |
| 44 | Ru | Ruthenium |
| 45 | Rh | Rhodium |
| 46 | Pd | Palladium |
| 47 | Ag | Silver |
| 48 | Cd | Cadmium |
| 49 | In | Indium |
| 50 | Sn | Tin |
| 51 | Sb | Antimony |
| 52 | Te | Tellurium |
| 53 | I | Iodine |
| 54 | Xe | Xenon |
| 55 | Cs | Cesium |
| 56 | Ba | Barium |
| 57 | La | Lanthanum |
| 58 | Ce | Cerium |
| 59 | Pr | Praseodymium |
| 60 | Nd | Neodymium |
| 61 | Pm | Promethium |
| 62 | Sm | Samarium |
| 63 | Eu | Europium |
| 64 | Gd | Gadolinium |
| 65 | Tb | Terbium |
| 66 | Dy | Dysprosium |
| 67 | Ho | Holmium |
| 68 | Er | Erbium |
| 69 | Tm | Thulium |
| 70 | Yb | Ytterbium |
| 71 | Lu | Lutetium |
| 72 | Hf | Hafnium |
| 73 | Ta | Tantalum |
| 74 | W | Tungsten |
| 75 | Re | Rhenium |
| 76 | Os | Osmium |
| 77 | Ir | Iridium |
| 78 | Pt | Platinum |
| 79 | Au | Gold |
| 80 | Hg | Mercury |
| 81 | Tl | Thallium |
| 82 | Pb | Lead |
| 83 | Bi | Bismuth |
| 84 | Po | Polonium |

*(continues)*

## ELEMENTS LISTED BY ATOMIC NUMBER *(continued)*

| | | |
|---|---|---|
| 85 | At | Astatine |
| 86 | Rn | Radon |
| 87 | Fr | Francium |
| 88 | Ra | Radium |
| 89 | Ac | Actinium |
| 90 | Th | Thorium |
| 91 | Pa | Protactinium |
| 92 | U | Uranium |
| 93 | Np | Neptunium |
| 94 | Pu | Plutonium |
| 95 | Am | Americium |
| 96 | Cm | Curium |
| 97 | Bk | Berkelium |
| 98 | Cf | Californium |
| 99 | Es | Einsteinium |
| 100 | Fm | Fermium |
| 101 | Md | Mendelevium |
| 102 | No | Nobelium |
| 103 | Lr | Lawrencium |
| 104 | Rf | Rutherfordium |
| 105 | Db | Dubnium |
| 106 | Sg | Seaborgium |
| 107 | Bh | Bohrium |
| 108 | Hs | Hassium |
| 109 | Mt | Meitnerium |
| 110 | Ds | Darmstadtium |
| 111 | Rg | Roentgenium |
| 112 | Cn | Copernicum |
| 113 | Uut | Ununtrium |
| 114 | Uuq | Ununquadium |
| 115 | Uup | Ununpentium |
| 116 | Uuh | Ununhexium |
| 117 | Uus | Ununseptium |
| 118 | Uuo | Ununoctium |

# Chronology

Civilization is essentially the story of the human mind understanding and gaining control over matter. The chronology presents some of the major milestones, scientific breakthroughs, and technical developments that formed the modern understanding of matter. Note that dates prior to 1543 are approximate.

**13.7 BILLION YEARS AGO** . . . . . Big bang event starts the universe.

**13.3 BILLION YEARS AGO** . . . . . The first stars form and begin to shine intensely.

**4.5 BILLION YEARS AGO** . . . . . . Earth forms within the primordial solar nebula.

**3.6 BILLION YEARS AGO** . . . . . . Life (simple microorganisms) appears in Earth's oceans.

**2,000,000–100,000 B.C.E.** . . . Early hunters of the Lower Paleolithic learn to use simple stone tools, such as handheld axes.

**100,000–40,000 B.C.E.** . . . . Neanderthal man of Middle Paleolithic lives in caves, controls fire, and uses improved stone tools for hunting.

**40,000–10,000 B.C.E.** . . . . . During the Upper Paleolithic, Cro-Magnon man displaces Neanderthal man. Cro-Magnon people develop more organized hunting and fishing activities using improved stone tools and weapons.

**8000–3500 B.C.E.** . . . . . . . . . Neolithic Revolution takes place in the ancient Middle East as people shift their dependence for subsistence from hunting and gathering to crop cultivation and animal domestication.

**3500–1200 B.C.E.** . . . . . . . . . . Bronze Age occurs in the ancient Middle East, when metalworking artisans start using bronze (a copper and tin alloy) to make weapons and tools.

**1200–600 B.C.E.** . . . . . . . . . . . People in the ancient Middle East enter the Iron Age. Eventually, the best weapons and tools are made of steel, an alloy of iron and varying amounts of carbon. The improved metal tools and weapons spread to Greece and later to Rome.

**1000 B.C.E.** .............. By this time, people in various ancient civilizations have discovered and are using the following chemical elements (in alphabetical order): carbon (C), copper (Cu), gold (Au), iron (Fe), lead (Pb), mercury (Hg), silver (Ag), sulfur (S), tin (Sn), and zinc (Zn).

**650 B.C.E.** ............... Kingdom of Lydia introduces officially minted gold and silver coins.

**600 B.C.E.** ............... Early Greek philosopher Thales of Miletus postulates that all substances come from water and would eventually turn back into water.

**450 B.C.E.** ............... Greek philosopher Empedocles proposes that all matter is made up of four basic elements (earth, air, water, and fire) that periodically combine and separate under the influence of two opposing forces (love and strife).

**430 B.C.E.** ............... Greek philosopher Democritus proposes that all things consist of changeless, indivisible, tiny pieces of matter called *atoms.*

**250 B.C.E.** ............... Archimedes of Syracuse designs an endless screw, later called the Archimedes screw. People use the fluid-moving device to remove water from the holds of sailing ships and to irrigate arid fields.

**300 C.E.** ................. Greek alchemist Zosimos of Panoplis writes the oldest known work describing alchemy.

**850** ..................... The Chinese use gunpowder for festive fireworks. It is a mixture of sulfur (S), charcoal (C), and potassium nitrate ($KNO_3$).

**1247** .................... British monk Roger Bacon writes the formula for gunpowder in his encyclopedic work *Opus Majus.*

**1250** .................... German theologian and natural philosopher Albertus Magnus isolates the element arsenic (As).

**1439** .................... Johannes Gutenberg successfully incorporates movable metal type in his mechanical printing press. His revolutionary approach to printing depends on a durable, hard metal alloy called type metal, which consists of a mixture of lead (Pb), tin (Sn), and antimony (Sb).

**1543** .................. Start of the Scientific Revolution. Polish astronomer Nicholas Copernicus promotes heliocentric (Sun-centered) cosmology with his deathbed publication of *On the Revolutions of Celestial Orbs.*

**1638** .................. Italian scientist Galileo Galilei publishes extensive work on solid mechanics, including uniform acceleration, free fall, and projectile motion.

**1643** .................. Italian physicist Evangelista Torricelli designs the first mercury barometer and then records the daily variation of atmospheric pressure.

**1661** .................. Irish-British scientist Robert Boyle publishes *The Sceptical Chymist,* in which he abandons the four classical Greek elements (earth, air, water, and fire) and questions how alchemists determine what substances are elements.

**1665** .................. British scientist Robert Hooke publishes *Micrographia,* in which he describes pioneering applications of the optical microscope in chemistry, botany, and other scientific fields.

**1667** .................. The work of German alchemist Johann Joachim Becher forms the basis of the phlogiston theory of heat.

**1669** .................. German alchemist Hennig Brand discovers the element phosphorous (P).

**1678** .................. Robert Hooke studies the action of springs and reports that the extension (or compression) of an elastic material takes place in direct proportion to the force exerted on the material.

**1687** .................. British physicist Sir Isaac Newton publishes *The Principia.* His work provides the mathematical foundations for understanding (from a classical physics perspective) the motion of almost everything in the physical universe.

1738 .................... Swiss mathematician Daniel Bernoulli publishes *Hydrodynamica*. In this seminal work, he identifies the relationships between density, pressure, and velocity in flowing fluids.

1748 .................... While conducting experiments with electricity, American statesman and scientist Benjamin Franklin coins the term *battery*.

1754 .................... Scottish chemist Joseph Black discovers a new gaseous substance, which he calls "fixed air." Other scientists later identify it as carbon dioxide ($CO_2$).

1764 .................... Scottish engineer James Watt greatly improves the Newcomen steam engine. Watt steam engines power the First Industrial Revolution.

1772 .................... Scottish physician and chemist Daniel Rutherford isolates a new colorless gaseous substance, calling it "noxious air." Other scientists soon refer to the new gas as nitrogen ($N_2$).

1785 .................... French scientist Charles-Augustin de Coulomb performs experiments that lead to the important law of electrostatics, later known as Coulomb's law.

1789 .................... French chemist Antoine-Laurent Lavoisier publishes *Treatise of Elementary Chemistry*, the first modern textbook on chemistry. Lavoisier also promotes the caloric theory of heat.

1800 .................... Italian physicist Count Alessandro Volta invents the voltaic pile. His device is the forerunner of the modern electric battery.

1803 .................... British schoolteacher and chemist John Dalton revives the atomic theory of matter. From his experiments, he concludes that all matter consists of combinations of atoms and that all the atoms of a particular element are identical.

1807 .................... British chemist Sir Humphry Davy discovers the element potassium (K) while experimenting with caustic potash (KOH). Potassium is the first metal isolated by the process of electrolysis.

1811 .................... Italian physicist Amedeo Avogadro proposes that equal volumes of different gases under the same conditions of pressure and temperature contain the same number of molecules. Scientists call this important hypothesis Avogadro's law.

1820 .................... Danish physicist Hans Christian Ørsted discovers a relationship between magnetism and electricity.

1824 .................... French military engineer Sadi Carnot publishes *Reflections on the Motive Power of Fire.* Despite the use of caloric theory, his work correctly identifies the general thermodynamic principles that govern the operation and efficiency of all heat engines.

1826 .................... French scientist André-Marie Ampère experimentally formulates the relationship between electricity and magnetism.

1827 .................... Experiments performed by German physicist George Simon Ohm indicate a fundamental relationship among voltage, current, and resistance.

1828 .................... Swedish chemist Jöns Jacob Berzelius discovers the element thorium (Th).

1831 .................... British experimental scientist Michael Faraday discovers the principle of electromagnetic induction. This principle is the basis for the electric dynamo.

Independent of Faraday, the American physicist Joseph Henry publishes a paper describing the electric motor (essentially a reverse dynamo).

1841 .................... German physicist and physician Julius Robert von Mayer states the conservation of energy principle, namely that energy can neither be created nor destroyed.

1847 .................... British physicist James Prescott Joule experimentally determines the mechanical equivalent of heat. Joule's work is a major step in developing the modern science of thermodynamics.

**1866** .................. Swedish scientist-industrialist Alfred Nobel finds a way to stabilize nitroglycerin and calls the new chemical explosive mixture dynamite.

**1869** .................. Russian chemist Dmitri Mendeleev introduces a periodic listing of the 63 known chemical elements in *Principles of Chemistry.* His periodic table includes gaps for elements predicted but not yet discovered.

American printer John W. Hyatt formulates celluloid, a flammable thermoplastic material made from a mixture of cellulose nitrate, alcohol, and camphor.

**1873** .................. Scottish mathematician and theoretical physicist James Clerk Maxwell publishes *Treatise on Electricity and Magnetism.*

**1876** .................. American physicist and chemist Josiah Willard Gibbs publishes *On the Equilibrium of Heterogeneous Substances.* This compendium forms the theoretical foundation of physical chemistry.

**1884** .................. Swedish chemist Svante Arrhenius proposes that electrolytes split or dissociate into electrically opposite positive and negative ions.

**1888** .................. German physicist Heinrich Rudolf Hertz produces and detects radio waves.

**1895** .................. German physicist Wilhelm Conrad Roentgen discovers X-rays.

**1896** .................. While investigating the properties of uranium salt, French physicist Antoine-Henri Becquerel discovers radioactivity.

**1897** .................. British physicist Sir Joseph John Thomson performs experiments that demonstrate the existence of the electron—the first subatomic particle discovered.

**1898** .................. French scientists Pierre and (Polish-born) Marie Curie announce the discovery of two new radioactive elements, polonium (Po) and radium (Ra).

**1900** .................... German physicist Max Planck postulates that blackbodies radiate energy only in discrete packets (or quanta) rather than continuously. His hypothesis marks the birth of quantum theory.

**1903** .................... New Zealand–born British physicist Baron (Ernest) Rutherford and British radiochemist Frederick Soddy propose the law of radioactive decay.

**1904** .................... German physicist Ludwig Prandtl revolutionizes fluid mechanics by introducing the concept of the boundary layer and its role in fluid flow.

**1905** .................... Swiss-German-American physicist Albert Einstein publishes the special theory of relativity, including the famous mass-energy equivalence formula ($E = mc^2$).

**1907** .................... Belgian-American chemist Leo Baekeland formulates bakelite. This synthetic thermoplastic material ushers in the age of plastics.

**1911**.................... Baron Ernest Rutherford proposes the concept of the atomic nucleus based on the startling results of an alpha particle–gold foil scattering experiment.

**1912** .................... German physicist Max von Laue discovers that X-rays are diffracted by crystals.

**1913** .................... Danish physicist Niels Bohr presents his theoretical model of the hydrogen atom—a brilliant combination of atomic theory with quantum physics.

Frederick Soddy proposes the existence of isotopes.

**1914** .................... British physicist Henry Moseley measures the characteristic X-ray lines of many chemical elements.

**1915** .................... Albert Einstein presents his general theory of relativity, which relates gravity to the curvature of space-time.

**1919** .................... Ernest Rutherford bombards nitrogen (N) nuclei with alpha particles, causing the nitrogen nuclei to transform into oxygen (O) nuclei and to emit protons (hydrogen nuclei).

British physicist Francis Aston uses the newly invented mass spectrograph to identify more than 200 naturally occurring isotopes.

**1923** .................. American physicist Arthur Holly Compton conducts experiments involving X-ray scattering that demonstrate the particle nature of energetic photons.

**1924** .................. French physicist Louis-Victor de Broglie proposes the particle-wave duality of matter.

**1926** .................. Austrian physicist Erwin Schrödinger develops quantum wave mechanics to describe the dual wave-particle nature of matter.

**1927** .................. German physicist Werner Heisenberg introduces his uncertainty principle.

**1929** .................. American astronomer Edwin Hubble announces that his observations of distant galaxies suggest an expanding universe.

**1932** .................. British physicist Sir James Chadwick discovers the neutron.

British physicist Sir John Cockcroft and Irish physicist Ernest Walton use a linear accelerator to bombard lithium (Li) with energetic protons, producing the first artificial disintegration of an atomic nucleus.

American physicist Carl D. Anderson discovers the positron.

**1934** ..................Italian-American physicist Enrico Fermi proposes a theory of beta decay that includes the neutrino. He also starts to bombard uranium with neutrons and discovers the phenomenon of slow neutrons.

**1938** .................. German chemists Otto Hahn and Fritz Strassmann bombard uranium with neutrons and detect the presence of lighter elements. Austrian physicist Lise Meitner and Austrian-British physicist Otto Frisch review Hahn's work and conclude in early 1939 that the German chemists had split the atomic nucleus, achieving neutron-induced nuclear fission.

E.I. du Pont de Nemours & Company introduces a new thermoplastic material called nylon.

**1941** .................. American nuclear scientist Glenn T. Seaborg and his associates use the cyclotron at the University of California, Berkeley, to synthesize plutonium (Pu).

**1942** .................. Modern nuclear age begins when Enrico Fermi's scientific team at the University of Chicago achieves the first self-sustained, neutron-induced fission chain reaction at Chicago Pile One (CP-1), a uranium-fueled, graphite-moderated atomic pile (reactor).

**1945** .................. American scientists successfully detonate the world's first nuclear explosion, a plutonium-implosion device code-named Trinity.

**1947** .................. American physicists John Bardeen, Walter Brattain, and William Shockley invent the transistor.

**1952** .................. A consortium of 11 founding countries establishes CERN, the European Organization for Nuclear Research, at a site near Geneva, Switzerland.

United States tests the world's first thermonuclear device (hydrogen bomb) at the Enewetak Atoll in the Pacific Ocean. Code-named Ivy Mike, the experimental device produces a yield of 10.4 megatons.

**1964** .................. German-American physicist Arno Allen Penzias and American physicist Robert Woodrow Wilson detect the cosmic microwave background (CMB).

**1967** .................. German-American physicist Hans Albrecht Bethe receives the 1967 Nobel Prize in physics for his theory of thermonuclear reactions being responsible for energy generation in stars.

**1969** .................. On July 20, American astronauts Neil Armstrong and Edwin "Buzz" Aldrin successfully land on the Moon as part of NASA's *Apollo 11* mission.

1972 .................. NASA launches the *Pioneer 10* spacecraft. It eventually becomes the first human-made object to leave the solar system on an interstellar trajectory

1985 .................. American chemists Robert F. Curl, Jr., and Richard E. Smalley, collaborating with British astronomer Sir Harold W. Kroto, discover the buckyball, an allotrope of pure carbon.

1996 .................. Scientists at CERN (near Geneva, Switzerland) announce the creation of antihydrogen, the first human-made antimatter atom.

1998 .................. Astrophysicists investigating very distant Type 1A supernovae discover that the universe is expanding at an accelerated rate. Scientists coin the term *dark energy* in their efforts to explain what these observations physically imply.

2001 .................. American physicist Eric A. Cornell, German physicist Wolfgang Ketterle, and American physicist Carl E. Wieman share the 2001 Nobel Prize in physics for their fundamental studies of the properties of Bose-Einstein condensates.

2005 .................. Scientists at the Lawrence Livermore National Laboratory (LLNL) in California and the Joint Institute for Nuclear Research (JINR) in Dubna, Russia, perform collaborative experiments that establish the existence of super-heavy element 118, provisionally called ununoctium (Uuo).

2008 .................. An international team of scientists inaugurates the world's most powerful particle accelerator, the Large Hadron Collider (LHC), located at the CERN laboratory near Geneva, Switzerland.

2009 .................. British scientist Charles Kao, American scientist Willard Boyle, and American scientist George Smith share the 2009 Nobel Prize in physics for their pioneering efforts in fiber optics and imaging semiconductor devices, developments that unleashed the information technology revolution.

**2010** .................. Element 112 is officially named Copernicum (Cn) by the IUPAC in honor of Polish astronomer Nicholas Copernicus (1473–1543), who championed heliocentric cosmology.

Scientists at the Joint Institute for Nuclear Research in Dubna, Russia, announce the synthesis of element 117 (ununseptium [Uus]) in early April.

# Glossary

**absolute zero** the lowest possible temperature; equal to 0 kelvin (K) (–459.67°F, –273.15°C)

**acceleration** (a) rate at which the velocity of an object changes with time

**accelerator** device for increasing the velocity and energy of charged elementary particles

**acid** substance that produces hydrogen ions ($H^+$) when dissolved in water

**actinoid** (formerly actinide) series of heavy metallic elements beginning with element 89 (actinium) and continuing through element 103 (lawrencium)

**activity** measure of the rate at which a material emits nuclear radiations

**air** overall mixture of gases that make up Earth's atmosphere

**alchemy** mystical blend of sorcery, religion, and prescientific chemistry practiced in many early societies around the world

**alloy** solid solution (compound) or homogeneous mixture of two or more elements, at least one of which is an elemental metal

**alpha particle** ($\alpha$) positively charged nuclear particle emitted from the nucleus of certain radioisotopes when they undergo decay; consists of two protons and two neutrons bound together

**alternating current** (AC) electric current that changes direction periodically in a circuit

**American customary system of units** (also American system) used primarily in the United States; based on the foot (ft), pound-mass (lbm), pound-force (lbf), and second (s). Peculiar to this system is the artificial construct (based on Newton's second law) that one pound-force equals one pound-mass (lbm) at sea level on Earth

**ampere** (A) SI unit of electric current

**anode** positive electrode in a battery, fuel cell, or electrolytic cell; oxidation occurs at anode

**antimatter** matter in which the ordinary nuclear particles are replaced by corresponding antiparticles

**Archimedes principle** the fluid mechanics rule that states that the buoyant (upward) force exerted on a solid object immersed in a fluid equals the weight of the fluid displaced by the object

**atom** smallest part of an element, indivisible by chemical means; consists of a dense inner core (nucleus) that contains protons and neutrons and a cloud of orbiting electrons

**atomic mass** *See* **relative atomic mass**

**atomic mass unit** (amu) 1/12 mass of carbon's most abundant isotope, namely carbon-12

**atomic number** (Z) total number of protons in the nucleus of an atom and its positive charge

**atomic weight** the mass of an atom relative to other atoms. *See also* **relative atomic mass**

**battery** electrochemical energy storage device that serves as a source of direct current or voltage

**becquerel** (Bq) SI unit of radioactivity; one disintegration (or spontaneous nuclear transformation) per second. *Compare with* **curie**

**beta particle** (β) elementary particle emitted from the nucleus during radioactive decay; a negatively charged beta particle is identical to an electron

**big bang** theory in cosmology concerning the origin of the universe; postulates that about 13.7 billion years ago, an initial singularity experienced a very large explosion that started space and time. Astrophysical observations support this theory and suggest that the universe has been expanding at different rates under the influence of gravity, dark matter, and dark energy

**blackbody** perfect emitter and perfect absorber of electromagnetic radiation; radiant energy emitted by a blackbody is a function only of the emitting object's absolute temperature

**black hole** incredibly compact, gravitationally collapsed mass from which nothing can escape

**boiling point** temperature (at a specified pressure) at which a liquid experiences a change of state into a gas

**Bose-Einstein condensate** (BEC) state of matter in which extremely cold atoms attain the same quantum state and behave essentially as a large "super atom"

**boson** general name given to any particle with a spin of an integral number (0, 1, 2, etc.) of quantum units of angular momentum. Carrier particles of all interactions are bosons. *See also* **carrier particle**

**brass** alloy of copper (Cu) and zinc (Zn)

**British thermal unit** (Btu) amount of heat needed to raise the temperature of 1 lbm of water 1°F at normal atmospheric pressure; 1 Btu = 1,055 J = 252 cal

**bronze** alloy of copper (Cu) and tin (Sn)

**calorie** (cal) quantity of heat; defined as the amount needed to raise one gram of water 1°C at normal atmospheric pressure; 1 cal = 4.1868 J = 0.004 Btu

**carbon dioxide** ($CO_2$) colorless, odorless, noncombustible gas present in Earth's atmosphere

**Carnot cycle** ideal reversible thermodynamic cycle for a theoretical heat engine; represents the best possible thermal efficiency of any heat engine operating between two absolute temperatures ($T_1$ and $T_2$)

**carrier particle** within the standard model, gluons are carrier particles for strong interactions; photons are carrier particles of electromagnetic interactions; and the W and Z bosons are carrier particles for weak interactions. *See also* **standard model**

**catalyst** substance that changes the rate of a chemical reaction without being consumed or changed by the reaction

**cathode** negative electrode in a battery, fuel cell, electrolytic cell, or electron (discharge) tube through which a primary stream of electrons enters a system

**chain reaction** reaction that stimulates its own repetition. *See also* **nuclear chain reaction**

**change of state** the change of a substance from one physical state to another; the atoms or molecules are structurally rearranged without experiencing a change in composition. Sometimes called change of phase or phase transition

**charged particle** elementary particle that carries a positive or negative electric charge

**chemical bond(s)** force(s) that holds atoms together to form stable configurations of molecules

**chemical property** characteristic of a substance that describes the manner in which the substance will undergo a reaction with another substance, resulting in a change in chemical composition. *Compare with* **physical property**

**chemical reaction** involves changes in the electron structure surrounding the nucleus of an atom; a dissociation, recombination, or rearrangement of atoms. During a chemical reaction, one or more kinds of matter (called reactants) are transformed into one or several new kinds of matter (called products)

**color charge** in the standard model, the charge associated with strong interactions. Quarks and gluons have color charge and thus participate in strong interactions. Leptons, photons, W bosons, and Z bosons do not have color charge and consequently do not participate in strong interactions. *See also* **standard model**

**combustion** chemical reaction (burning or rapid oxidation) between a fuel and oxygen that generates heat and usually light

**composite materials** human-made materials that combine desirable properties of several materials to achieve an improved substance; includes combinations of metals, ceramics, and plastics with built-in strengthening agents

**compound** pure substance made up of two or more elements chemically combined in fixed proportions

**compressible flow** fluid flow in which density changes cannot be neglected

**compression** condition when an applied external force squeezes the atoms of a material closer together. *Compare with* **tension**

**concentration** for a solution, the quantity of dissolved substance per unit quantity of solvent

**condensation** change of state process by which a vapor (gas) becomes a liquid. *The opposite of* **evaporation**

**conduction** (thermal) transport of heat through an object by means of a temperature difference from a region of higher temperature to a region of lower temperature. *Compare with* **convection**

**conservation of mass and energy** Einstein's special relativity principle stating that energy (E) and mass (m) can neither be created nor destroyed, but are interchangeable in accordance with the equation $E = mc^2$, where $c$ represents the speed of light

**convection** fundamental form of heat transfer characterized by mass motions within a fluid resulting in the transport and mixing of the properties of that fluid

**coulomb** (C) SI unit of electric charge; equivalent to quantity of electric charge transported in one second by a current of one ampere

**covalent bond** the chemical bond created within a molecule when two or more atoms share an electron

**creep** slow, continuous, permanent deformation of solid material caused by a constant tensile or compressive load that is less than the load necessary for the material to give way (yield) under pressure. *See also* **plastic deformation**

**crystal** a solid whose atoms are arranged in an orderly manner, forming a distinct, repetitive pattern

**curie** (Ci) traditional unit of radioactivity equal to 37 billion ($37 \times 10^9$) disintegrations per second. *Compare with* **becquerel**

**current** (I) flow of electric charge through a conductor

**dark energy** a mysterious natural phenomenon or unknown cosmic force thought responsible for the observed acceleration in the rate of expansion of the universe. Astronomical observations suggest dark energy makes up about 72 percent of the universe

**dark matter** (nonbaryonic matter) exotic form of matter that emits very little or no electromagnetic radiation. It experiences no measurable interaction with ordinary (baryonic) matter but somehow accounts for the observed structure of the universe. It makes up about 23 percent of the content of the universe, while ordinary matter makes up less than 5 percent

**density** ($\rho$) mass of a substance per unit volume at a specified temperature

**deposition** direct transition of a material from the gaseous (vapor) state to the solid state without passing through the liquid phase. *Compare with* **sublimation**

**dipole magnet** any magnet with one north and one south pole

**direct current** (DC) electric current that always flows in the same direction through a circuit

**elastic deformation** temporary change in size or shape of a solid due to an applied force (stress); when force is removed the solid returns to its original size and shape

**elasticity** ability of a body that has been deformed by an applied force to return to its original shape when the force is removed

**elastic modulus** a measure of the stiffness of a solid material; defined as the ratio of stress to strain

**electricity** flow of energy due to the motion of electric charges; any physical effect that results from the existence of moving or stationary electric charges

**electrode** conductor (terminal) at which electricity passes from one medium into another; positive electrode is the *anode;* negative electrode is the *cathode*

**electrolyte** a chemical compound that, in an aqueous (water) solution, conducts an electric current

**electromagnetic radiation** (EMR) oscillating electric and magnetic fields that propagate at the speed of light. Includes in order of increasing frequency and energy: radio waves, radar waves, infrared (IR) radiation, visible light, ultraviolet radiation, X-rays, and gamma rays

**electron** (e) stable elementary particle with a unit negative electric charge ($1.602 \times 10^{-19}$ C). Electrons form an orbiting cloud, or shell, around the positively charged atomic nucleus and determine an atom's chemical properties

**electron volt** (eV) energy gained by an electron as it passes through a potential difference of one volt; one electron volt has an energy equivalence of $1.519 \times 10^{-22}$ Btu = $1.602 \times 10^{-19}$ J

**element** pure chemical substance indivisible into simpler substances by chemical means; all the atoms of an element have the same number of protons in the nucleus and the same number of orbiting electrons, although the number of neutrons in the nucleus may vary

**elementary particle** a fundamental constituent of matter; the basic atomic model suggests three elementary particles: the proton, neutron, and electron. *See also* **fundamental particle**

**endothermic reaction** chemical reaction requiring an input of energy to take place. *Compare with* **exothermic reaction**

**energy** (E) capacity to do work; appears in many different forms, such as mechanical, thermal, electrical, chemical, and nuclear

**entropy** (S) measure of disorder within a system; as entropy increases, energy becomes less available to perform useful work

**evaporation** physical process by which a liquid is transformed into a gas (vapor) at a temperature below the boiling point of the liquid. *Compare with* **sublimation**

**excited state** state of a molecule, atom, electron, or nucleus when it possesses more than its normal energy. *Compare with* **ground state**

**exothermic reaction** chemical reaction that releases energy as it takes place. *Compare with* **endothermic reaction**

**fatigue** weakening or deterioration of metal or other material that occurs under load, especially under repeated cyclic or continued loading

**fermion** general name scientists give to a particle that is a matter constituent. Fermions are characterized by spin in odd half-integer quantum units (namely, 1/2, 3/2, 5/2, etc.); quarks, leptons, and baryons are all fermions

**fission** (nuclear) splitting of the nucleus of a heavy atom into two lighter nuclei accompanied by the release of a large amount of energy as well as neutrons, X-rays, and gamma rays

**flavor** in the standard model, quantum number that distinguishes different types of quarks and leptons. *See also* **quark; lepton**

**fluid mechanics** scientific discipline that deals with the behavior of fluids (both gases and liquids) at rest (fluid statics) and in motion (fluid dynamics)

**foot-pound (force)** ($ft\text{-}lb_{force}$) unit of work in American customary system of units; 1 $ft\text{-}lb_{force}$ = 1.3558 J

**force** (F) the cause of the acceleration of material objects as measured by the rate of change of momentum produced on a free body. Force is a vector quantity mathematically expressed by Newton's second law of motion: force = mass × acceleration

**freezing point** the temperature at which a substance experiences a change from the liquid state to the solid state at a specified pressure; at this temperature, the solid and liquid states of a substance can coexist in equilibrium. *Synonymous with* **melting point**

**fundamental particle** particle with no internal substructure; in the standard model, any of the six types of quarks or six types of leptons and their antiparticles. Scientists postulate that all other particles are made from a combination of quarks and leptons. *See also* **elementary particle**

**fusion** (nuclear) nuclear reaction in which lighter atomic nuclei join together (fuse) to form a heavier nucleus, liberating a great deal of energy

***g*** acceleration due to gravity at sea level on Earth; approximately 32.2 $ft/s^2$ (9.8 $m/s^2$)

**gamma ray** ($\gamma$) high-energy, very short–wavelength photon of electromagnetic radiation emitted by a nucleus during certain nuclear reactions or radioactive decay

**gas** state of matter characterized as an easily compressible fluid that has neither a constant volume nor a fixed shape; a gas assumes the total size and shape of its container

**gravitational lensing** bending of light from a distant celestial object by a massive (gravitationally influential) foreground object

**ground state** state of a nucleus, atom, or molecule at its lowest (normal) energy level

**hadron** any particle (such as a baryon) that exists within the nucleus of an atom; made up of quarks and gluons, hadrons interact with the strong force

**half-life** (radiological) time in which half the atoms of a particular radioactive isotope disintegrate to another nuclear form

**heat** energy transferred by a temperature difference or thermal process. *Compare* **work**

**heat capacity** (c) amount of heat needed to raise the temperature of an object by one degree

**heat engine** thermodynamic system that receives energy in the form of heat and that, in the performance of energy transformation on a working fluid, does work. Heat engines function in thermodynamic cycles

**hertz** (Hz) SI unit of frequency; equal to one cycle per second

**high explosive** (HE) energetic material that detonates (rather than burns); the rate of advance of the reaction zone into the unreacted material exceeds the velocity of sound in the unreacted material

**horsepower** (hp) American customary system unit of power; 1 hp = 550 $ft\text{-}lb_{force}/s$ = 746 W

**hydraulic** operated, moved, or affected by liquid used to transmit energy

**hydrocarbon** organic compound composed of only carbon and hydrogen atoms

**ideal fluid** *See* **perfect fluid**

**ideal gas law** important physical principle: $P\,V = n\,R_u\,T$, where P is pressure, V is volume, T is temperature, $n$ is the number of moles of gas, and $R_u$ is the universal gas constant

**incompressible flow** fluid flow in which density changes can be neglected. *Compare with* **compressible flow**

**inertia** resistance of a body to a change in its state of motion

**infrared (IR) radiation** that portion of the electromagnetic (EM) spectrum lying between the optical (visible) and radio wavelengths

**International System of units** *See* **SI unit system**

**inviscid fluid** perfect fluid that has zero coefficient of viscosity

**ion** atom or molecule that has lost or gained one or more electrons, so that the total number of electrons does not equal the number of protons

**ionic bond** formed when one atom gives up at least one outer electron to another atom, creating a chemical bond–producing electrical attraction between the atoms

**isotope** atoms of the same chemical element but with different numbers of neutrons in their nucleus

**joule** (J) basic unit of energy or work in the SI unit system; 1 J = 0.2388 calorie = 0.00095 Btu

**kelvin** (K) SI unit of absolute thermodynamic temperature

**kinetic energy** (KE) energy due to motion

**lepton** fundamental particle of matter that does not participate in strong interactions; in the standard model, the three charged leptons are the electron (e), the muon ($\mu$), and the tau ($\tau$) particle; the three neutral leptons are the electron neutrino ($\nu_e$), the muon neutrino ($\nu_\mu$), and the tau neutrino ($\nu_\tau$). A corresponding set of antiparticles also exists. *See also* **standard model**

**light-year** (ly) distance light travels in one year; 1 ly $\approx 5.88 \times 10^{12}$ miles ($9.46 \times 10^{12}$ km)

**liquid** state of matter characterized as a relatively incompressible flowing fluid that maintains an essentially constant volume but assumes the shape of its container

**liter** (l or L) SI unit of volume; 1 L = 0.264 gal

**magnet** material or device that exhibits magnetic properties capable of causing the attraction or repulsion of another magnet or the attraction of certain ferromagnetic materials such as iron

**manufacturing** process of transforming raw material(s) into a finished product, especially in large quantities

**mass** (m) property that describes how much material makes up an object and gives rise to an object's inertia

**mass number** *See* **relative atomic mass**

**mass spectrometer** instrument that measures relative atomic masses and relative abundances of isotopes

**material** tangible substance (chemical, biological, or mixed) that goes into the makeup of a physical object

**mechanics** branch of physics that deals with the motions of objects

**melting point** temperature at which a substance experiences a change from the solid state to the liquid state at a specified pressure; at this temperature, the solid and liquid states of a substance can coexist in equilibrium. *Synonymous with* **freezing point**

**metallic bond** chemical bond created as many atoms of a metallic substance share the same electrons

**meter** (m) fundamental SI unit of length; 1 meter = 3.281 feet. British spelling *metre*

**metric system** *See* **SI unit system**

**metrology** science of dimensional measurement; sometimes includes the science of weighing

**microwave** (radiation) comparatively short-wavelength electromagnetic (EM) wave in the radio frequency portion of the EM spectrum

**mirror matter** *See* **antimatter**

**mixture** a combination of two or more substances, each of which retains its own chemical identity

**molarity** (M) concentration of a solution expressed as moles of solute per kilogram of solvent

**mole** (mol) SI unit of the amount of a substance; defined as the amount of substance that contains as many elementary units as there are atoms in 0.012 kilograms of carbon-12, a quantity known as Avogadro's number ($N_A$), which has a value of about $6.022 \times 10^{23}$ molecules/mole

**molecule** smallest amount of a substance that retains the chemical properties of the substance; held together by chemical bonds, a molecule can consist of identical atoms or different types of atoms

**monomer** substance of relatively low molecular mass; any of the small molecules that are linked together by covalent bonds to form a polymer

**natural material** material found in nature, such as wood, stone, gases, and clay

**neutrino** ($\nu$) lepton with no electric charge and extremely low (if not zero) mass; three known types of neutrinos are the electron neutrino ($\nu_e$), the muon neutrino ($\nu_\mu$), and the tau neutrino ($\nu_\tau$). *See also* **lepton**

**neutron** (n) an uncharged elementary particle found in the nucleus of all atoms except ordinary hydrogen. Within the standard model, the neutron is a baryon with zero electric charge consisting of two down (d) quarks and one up (u) quark. *See also* **standard model**

**newton** (N) The SI unit of force; 1 N = 0.2248 lbf

**nuclear chain reaction** occurs when a fissionable nuclide (such as plutonium-239) absorbs a neutron, splits (or fissions), and releases several neutrons along with energy. A fission chain reaction is self-sustaining when (on average) at least one released neutron per fission event survives to create another fission reaction

**nuclear energy** energy released by a nuclear reaction (fission or fusion) or by radioactive decay

**nuclear radiation** particle and electromagnetic radiation emitted from atomic nuclei as a result of various nuclear processes, such as radioactive decay and fission

**nuclear reaction** reaction involving a change in an atomic nucleus, such as fission, fusion, neutron capture, or radioactive decay

**nuclear reactor** device in which a fission chain reaction can be initiated, maintained, and controlled

**nuclear weapon** precisely engineered device that releases nuclear energy in an explosive manner as a result of nuclear reactions involving fission, fusion, or both

**nucleon** constituent of an atomic nucleus; a proton or a neutron

**nucleus** (plural: nuclei) small, positively charged central region of an atom that contains essentially all of its mass. All nuclei contain both protons and neutrons except the nucleus of ordinary hydrogen, which consists of a single proton

**nuclide** general term applicable to all atomic (isotopic) forms of all the elements; nuclides are distinguished by their atomic number, relative mass number (atomic mass), and energy state

**ohm** (Ω) SI unit of electrical resistance

**oxidation** chemical reaction in which oxygen combines with another substance, and the substance experiences one of three processes: (1) the gaining of oxygen, (2) the loss of hydrogen, or (3) the loss of electrons. In these reactions, the substance being "oxidized" loses electrons and forms positive ions. *Compare with* **reduction**

**oxidation-reduction (redox) reaction** chemical reaction in which electrons are transferred between species or in which atoms change oxidation number

**particle** minute constituent of matter, generally one with a measurable mass

**pascal** (Pa) SI unit of pressure; 1 Pa = 1 $N/m^2$ = 0.000145 psi

**Pascal's principle** when an enclosed (static) fluid experiences an increase in pressure, the increase is transmitted throughout the fluid; the physical principle behind all hydraulic systems

**Pauli exclusion principle** postulate that no two electrons in an atom can occupy the same quantum state at the same time; also applies to protons and neutrons

**perfect fluid** hypothesized fluid primarily characterized by a lack of viscosity and usually by incompressibility

**perfect gas law** *See* **ideal gas law**

**periodic table** list of all the known elements, arranged in rows (periods) in order of increasing atomic numbers and columns (groups) by similar physical and chemical characteristics

**phase** one of several different homogeneous materials present in a portion of matter under study; the set of states of a large-scale (macroscopic) physical system having relatively uniform physical properties and chemical composition

**phase transition** *See* **change of state**

**photon** A unit (or particle) of electromagnetic radiation that carries a quantum (packet) of energy that is characteristic of the particular radiation. Photons travel at the speed of light and have an effective momentum, but no mass or electrical charge. In the standard model, a photon is the carrier particle of electromagnetic radiation

**photovoltaic cell** *See* **solar cell**

**physical property** characteristic quality of a substance that can be measured or demonstrated without changing the composition or chemical identity of the substance, such as temperature and density. *Compare with* **chemical property**

**Planck's constant** (h) fundamental physical constant describing the extent to which quantum mechanical behavior influences nature. Equals the ratio of a photon's energy (E) to its frequency (ν), namely: $h = E/\nu = 6.626 \times 10^{-34}$ J-s ($6.282 \times 10^{-37}$ Btu-s). *See also* **uncertainty principle**

**plasma** electrically neutral gaseous mixture of positive and negative ions; called the fourth state of matter

**plastic deformation** permanent change in size or shape of a solid due to an applied force (stress)

**plasticity** tendency of a loaded body to assume a (deformed) state other than its original state when the load is removed

**plastics** synthesized family of organic (mainly hydrocarbon) polymer materials used in nearly every aspect of modern life

**pneumatic** operated, moved, or effected by a pressurized gas (typically air) that is used to transmit energy

**polymer** very large molecule consisting of a number of smaller molecules linked together repeatedly by covalent bonds, thereby forming long chains

**positron** ($e^+$ or $\beta^+$) elementary antimatter particle with the mass of an electron but charged positively

**pound-force** (lbf) basic unit of force in the American customary system; 1 lbf = 4.448 N

**pound-mass** (lbm) basic unit of mass in the American customary system; 1 lbm = 0.4536 kg

**power** rate with respect to time at which work is done or energy is transformed or transferred to another location; 1 hp = 550 ft-$lb_{force}$/s = 746 W

**pressure** (P) the normal component of force per unit area exerted by a fluid on a boundary; 1 psi = 6,895 Pa

**product** substance produced by or resulting from a chemical reaction

**proton** (p) stable elementary particle with a single positive charge. In the the standard model, the proton is a baryon with an electric charge of +1; it consists of two up (u) quarks and one down (d) quark. *See also* **standard model**

**quantum mechanics** branch of physics that deals with matter and energy on a very small scale; physical quantities are restricted to discrete values and energy to discrete packets called quanta

**quark** fundamental matter particle that experiences strong-force interactions. The six flavors of quarks in order of increasing mass are up (u), down (d), strange (s), charm (c), bottom (b), and top (t)

**radiation heat transfer** The transfer of heat by electromagnetic radiation that arises due to the temperature of a body; can takes place in and through a vacuum

**radioactive isotope** unstable isotope of an element that decays or disintegrates spontaneously, emitting nuclear radiation; also called radioisotope

**radioactivity** spontaneous decay of an unstable atomic nucleus, usually accompanied by the emission of nuclear radiation, such as alpha particles, beta particles, gamma rays, or neutrons

**radio frequency** (RF) a frequency at which electromagnetic radiation is useful for communication purposes; specifically, a frequency above 10,000 hertz (Hz) and below $3 \times 10^{11}$Hz

**rankine** (R) American customary unit of absolute temperature. *See also* **kelvin (K)**

**reactant** original substance or initial material in a chemical reaction

**reduction** portion of an oxidation-reduction (redox) reaction in which there is a gain of electrons, a gain in hydrogen, or a loss of oxygen. *See also* **oxidation-reduction (redox) reaction**

**relative atomic mass** (A) total number of protons and neutrons (nucleons) in the nucleus of an atom. Previously called *atomic mass* or *atomic mass number. See also* **atomic mass unit**

**residual electromagnetic effect** force between electrically neutral atoms that leads to the formation of molecules

**residual strong interaction** interaction responsible for the nuclear binding force—that is, the strong force holding hadrons (protons and neutrons) together in the atomic nucleus. *See also* **strong force**

**resilience** property of a material that enables it to return to its original shape and size after deformation

**resistance** (R) the ratio of the voltage (V) across a conductor to the electric current (I) flowing through it

**scientific notation** A method of expressing powers of 10 that greatly simplifies writing large numbers; for example, $3 \times 10^6 = 3,000,000$

**SI unit system** international system of units (the metric system), based upon the meter (m), kilogram (kg), and second (s) as the fundamental units of length, mass, and time, respectively

**solar cell** (photovoltaic cell) a semiconductor direct energy conversion device that transforms sunlight into electric energy

**solid** state of matter characterized by a three-dimensional regularity of structure; a solid is relatively incompressible, maintains a fixed volume, and has a definitive shape

**solution** When scientists dissolve a substance in a pure liquid, they refer to the dissolved substance as the *solute* and the host pure liquid as the *solvent*. They call the resulting intimate mixture the solution

**spectroscopy** study of spectral lines from various atoms and molecules; emission spectroscopy infers the material composition of the objects that emitted the light; absorption spectroscopy infers the composition of the intervening medium

**speed of light** *(c)* speed at which electromagnetic radiation moves through a vacuum; regarded as a universal constant equal to 186,283.397 mi/s (299,792.458 km/s)

**stable isotope** isotope that does not undergo radioactive decay

**standard model** contemporary theory of matter, consisting of 12 fundamental particles (six quarks and six leptons), their respective antiparticles, and four force carriers (gluons, photons, W bosons, and Z bosons)

**state of matter** form of matter having physical properties that are quantitatively and qualitatively different from other states of matter; the three more common states on Earth are solid, liquid, and gas

**steady state** condition of a physical system in which parameters of importance (fluid velocity, temperature, pressure, etc.) do not vary significantly with time

**strain** the change in the shape or dimensions (volume) of an object due to applied forces; longitudinal, volume, and shear are the three basic types of strain

**stress** applied force per unit area that causes an object to deform (experience strain); the three basic types of stress are compressive (or tensile) stress, hydrostatic pressure, and shear stress

**string theory** theory of quantum gravity that incorporates Einstein's general relativity with quantum mechanics in an effort to explain space-time phenomena on the smallest imaginable scales; vibrations of incredibly tiny stringlike structures form quarks and leptons

**strong force** In the standard model, the fundamental force between quarks and gluons that makes them combine to form hadrons, such as protons and neutrons; also holds hadrons together in a nucleus. *See also* **standard model**

**subatomic particle** any particle that is small compared to the size of an atom

**sublimation** direct transition of a material from the solid state to the gaseous (vapor) state without passing through the liquid phase. *Compare with* **deposition**

**superconductivity** the ability of a material to conduct electricity without resistance at a temperature above absolute zero

**temperature** (T) thermodynamic property that serves as a macroscopic measure of atomic and molecular motions within a substance; heat naturally flows from regions of higher temperature to regions of lower temperature

**tension** condition when applied external forces pull atoms of a material farther apart. *Compare with* **compression**

**thermal conductivity** (k) intrinsic physical property of a substance; a material's ability to conduct heat as a consequence of molecular motion

**thermodynamics** branch of science that treats the relationships between heat and energy, especially mechanical energy

**thermodynamic system** collection of matter and space with boundaries defined in such a way that energy transfer (as work and heat) from and to the system across these boundaries can be easily identified and analyzed

**thermometer** instrument or device for measuring temperature

**toughness** ability of a material (especially a metal) to absorb energy and deform plastically before fracturing

**transmutation** transformation of one chemical element into a different chemical element by a nuclear reaction or series of reactions

**transuranic element** (isotope) human-made element (isotope) beyond uranium on the periodic table

**ultraviolet (UV) radiation** portion of the electromagnetic spectrum that lies between visible light and X-rays

**uncertainty principle** Heisenberg's postulate that places quantum-level limits on how accurately a particle's momentum $(p)$ and position $(x)$ can be simultaneously measured. Planck's constant (h) expresses this uncertainty as $\Delta x \times \Delta p \geq h/2\pi$

**U.S. customary system of units** *See* **American customary system of units**

**vacuum** relative term used to indicate the absence of gas or a region in which there is a very low gas pressure

**valence electron** electron in the outermost shell of an atom

**van der Waals force** generally weak interatomic or intermolecular force caused by polarization of electrically neutral atoms or molecules

**vapor** gaseous state of a substance

**velocity** vector quantity describing the rate of change of position; expressed as length per unit of time

**velocity of light** ($c$) *See* **speed of light**

**viscosity** measure of the internal friction or flow resistance of a fluid when it is subjected to shear stress

**volatile** solid or liquid material that easily vaporizes; volatile material has a relatively high vapor pressure at normal temperatures

**volt** (V) SI unit of electric potential difference

**volume** (V) space occupied by a solid object or a mass of fluid (liquid or confined gas)

**watt** (W) SI unit of power (work per unit time); 1 W = 1 J/s = 0.00134 hp = 0.737 $\text{ft-lb}_{\text{force}}/\text{s}$

**wavelength** ($\lambda$) the mean distance between two adjacent maxima (or minima) of a wave

**weak force** fundamental force of nature responsible for various types of radioactive decay

**weight** ($w$) the force of gravity on a body; on Earth, product of the mass (m) of a body times the acceleration of gravity ($g$), namely $w = \text{m} \times g$

**work** (W) energy expended by a force acting though a distance. *Compare with* **heat**

**X-ray** penetrating form of electromagnetic (EM) radiation that occurs on the EM spectrum between ultraviolet radiation and gamma rays

# Further Resources

## BOOKS

Allcock, Harry R. *Introduction to Materials Chemistry.* New York: John Wiley & Sons, 2008. A college-level textbook that provides a basic treatment of the principles of chemistry upon which materials science depends.

Angelo, Joseph A., Jr. *Nuclear Technology.* Westport, Conn.: Greenwood Press, 2004. The book provides a detailed discussion of both military and civilian nuclear technology and includes impacts, issues, and future advances.

———. *Encyclopedia of Space and Astronomy.* New York: Facts On File, 2006. Provides a comprehensive treatment of major concepts in astronomy, astrophysics, planetary science, cosmology, and space technology.

Ball, Philip. *Designing the Molecular World: Chemistry at the Frontier.* Princeton, N.J.: Princeton University Press, 1996. Discusses many recent advances in modern chemistry, including nanotechnology and superconductor materials.

———. *Made to Measure: New Materials for the 21st Century.* Princeton, N.J.: Princeton University Press, 1998. Discusses how advanced new materials can significantly influence life in the 21st century.

Bensaude-Vincent, Bernadette, and Isabelle Stengers. *A History of Chemistry.* Cambridge, Mass.: Harvard University Press, 1996. Describes how chemistry emerged as a science and its impact on civilization.

Callister, William D., Jr. *Materials Science and Engineering: An Introduction.* 8th ed. New York: John Wiley & Sons, 2010. Intended primarily for engineers, technically knowledgeable readers will also benefit from this book's introductory treatment of metals, ceramics, polymers, and composite materials.

Charap, John M. *Explaining the Universe: The New Age of Physics.* Princeton, N.J.: Princeton University Press, 2004. Discusses the important discoveries in physics during the 20th century that are influencing civilization.

Close, Frank, et al. *The Particle Odyssey: A Journey to the Heart of the Matter.* New York: Oxford University Press, 2002. A well-illustrated and enjoyable tour of the subatomic world.

Cobb, Cathy, and Harold Goldwhite. *Creations of Fire: Chemistry's Lively History from Alchemy to the Atomic Age.* New York: Plenum Press, 1995. Uses historic circumstances and interesting individuals to describe the emergence of chemistry as a scientific discipline.

Feynman, Richard P. *QED: The Strange Theory of Light and Matter.* Princeton, N.J.: Princeton University Press, 2006. Written by an American Nobel laureate, addresses several key topics in modern physics.

Gordon, J. E. *The New Science of Strong Materials or Why You Don't Fall Through the Floor.* Princeton, N.J.: Princeton University Press, 2006. Discusses the science of structural materials in a manner suitable for both technical and lay audiences.

Hill, John W., and Doris K. Kolb. *Chemistry for Changing Times.* 11th ed. Upper Saddle River, N.J.: Pearson Prentice Hall, 2007. Readable college-level textbook that introduces all the basic areas of modern chemistry.

Krebs, Robert E. *The History and Use of Our Earth's Chemical Elements: A Reference Guide.* 2nd ed. Westport, Conn.: Greenwood Press, 2006. Provides a concise treatment of each of the chemical elements.

Levere, Trevor H. *Transforming Matter: A History of Chemistry from Alchemy to the Buckyball.* Baltimore: Johns Hopkins University Press, 2001. Provides an understandable overview of the chemical sciences from the early alchemists through modern times.

Lutgens, Frederick K., and Edward J. Tarbuck. *The Atmosphere: An Introduction to Meteorology.* 10th ed. Upper Saddle River, N.J.: Pearson Prentice Hall, 2007. Readable college-level textbook that discusses the atmosphere, meteorology, climate, and the physical properties of air.

Mackintosh, Ray, et al. *Nucleus: A Trip into the Heart of Matter.* Baltimore: Johns Hopkins University Press, 2001. Provides a technical though readable explanation of how modern scientists developed their current understanding of the atomic nucleus and the standard model.

Nicolaou, K. C., and Tamsyn Montagnon. *Molecules that Changed the World.* New York: John Wiley & Sons, 2008. Provides an interesting treatment of such important molecules as aspirin, camphor, glucose, quinine, and morphine.

Scerri, Eric R. *The Periodic Table: Its Story and Its Significance.* New York: Oxford University Press, 2007. Provides a detailed look at the periodic table and its iconic role in the practice of modern science.

Smith, William F., and Javad Hashemi. *Foundations of Materials Science and Engineering.* 5th ed. New York: McGraw-Hill, 2006. Provides scientists and engineers of all disciplines an introduction to materials science, including metals, ceramics, polymers, and composite materials. Technically knowledgeable laypersons will find the treatment of specific topics such as biological materials useful.

Strathern, Paul. *Mendeleyev's Dream: The Quest for the Elements.* New York: St. Martin's Press, 2001. Describes the intriguing history of chemistry from the early Greek philosophers to the 19th-century Russian chemist Dmitri Mendeleyev.

Thrower, Peter, and Thomas Mason. *Materials in Today's World.* 3rd ed. New York: McGraw-Hill Companies, 2007. Provides a readable introductory treatment of modern materials science, including biomaterials and nanomaterials.

Trefil, James, and Robert M. Hazen. *Physics Matters: An Introduction to Conceptual Physics.* New York: John Wiley & Sons, 2004. Highly-readable introductory college-level textbook that provides a good overview of physics from classical mechanics to relativity and cosmology. Laypersons will find the treatment of specific topics useful and comprehendible.

Zee, Anthony. *Quantum Field Theory in a Nutshell.* Princeton, N.J.: Princeton University Press, 2003. A reader-friendly treatment of the generally complex and profound physical concepts that constitute quantum field theory.

## WEB SITES

To help enrich the content of this book and to make your investigation of matter more enjoyable, the following is a selective list of recommended Web sites. Many of the sites below will also lead to other interesting science-related locations on the Internet. Some sites provide unusual science learning opportunities (such as laboratory simulations) or in-depth educational resources.

**American Chemical Society (ACS)** is a congressionally chartered independent membership organization that represents professionals at all degree levels and in all fields of science involving chemistry. The ACS Web site includes educational resources for high school and college students. Available online. URL: http://portal.acs.org/portal/acs/corg/content. Accessed on February 12, 2010.

**American Institute of Physics (AIP)** is a not-for-profit corporation that promotes the advancement and diffusion of the knowledge of physics and its applications to human welfare. This Web site offers an enormous quantity of fascinating information about the history of physics from ancient Greece up to the present day. Available online. URL: http://www.aip.org/aip/. Accessed on February 12, 2010.

**Chandra X-ray Observatory (CXO)** is a space-based NASA astronomical observatory that observes the universe in the X-ray portion of the electromagnetic spectrum. This Web site contains contemporary information and educational materials about astronomy, astrophysics, and cosmology, including topics such as black holes, neutron stars, dark matter, and dark energy. Available online. URL: http://www.chandra.harvard.edu/. Accessed on February 12, 2010.

**The ChemCollective** is an online resource for learning about chemistry. Through simulations developed by the Department of Chemistry of Carnegie Mellon University (with funding from the National Science Foundation), a person gets the chance to safely mix chemicals without worrying about accidentally spilling them. Available online. URL: http://www.chemcollective.org/vlab/vlab.php. Accessed on February 12, 2010.

**Chemical Heritage Foundation (CHF)** maintains a rich and informative collection of materials that describe the history and heritage of the chemical and molecular sciences, technologies, and industries. Available online. URL: http://www.chemheritage.org/. Accessed on February 12, 2010.

**Department of Defense (DOD)** is responsible for maintaining armed forces of sufficient strength and technology to protect the United States and its citizens from all credible foreign threats. This Web site serves as an efficient access point to activities within the DOD, including those taking place within each of the individual armed services: the U.S. Army, U.S. Navy, U.S. Air Force, and U.S. Marines. As part of national security, the DOD sponsors a large amount of research and development, including activities in materials science, chemistry, physics, and nanotechnology. Available online. URL: http://www.defenselink.mil/. Accessed on February 12, 2010.

**Department of Energy (DOE)** is the single largest supporter of basic research in the physical sciences in the federal government of the United States. Topics found on this Web site include materials sciences, nanotechnology, energy sciences, chemical science, high-energy physics, and nuclear physics. The Web site also includes convenient links to all of the DOE's national laboratories. Available online. URL: http://energy.gov/. Accessed on February 12, 2010.

**Fermi National Accelerator Laboratory (Fermilab)** performs research that advances the understanding of the fundamental nature of matter and energy. Fermilab's Web site contains contemporary information about particle physics, the standard model, and the impact of particle physics on society. Available online. URL: http://www.fnal.gov/. Accessed on February 12, 2010.

***Hubble Space Telescope (HST)*** is a space-based NASA observatory that has examined the universe in the (mainly) visible portion of the electromagnetic spectrum. This Web site contains contemporary information and educational materials about astronomy, astrophysics, and cosmology, including topics such as black holes, neutron stars, dark matter, and dark energy. Available online. URL: http://hubblesite.org/. Accessed on February 12, 2010.

**Institute and Museum of the History of Science** in Florence, Italy, offers a special collection of scientific instruments (some viewable online), including those used by Galileo Galilei. Available online. URL: http://www.imss.fi.it/. Accessed on February 12, 2010.

**International Union of Pure and Applied Chemistry (IUPAC)** is an international nongovernmental organization that fosters worldwide communications in the chemical sciences and in providing a common language for chemistry that unifies the industrial, academic, and public sectors. Available online. URL: http://www.iupac.org/. Accessed on February 12, 2010.

**National Aeronautics and Space Administration (NASA)** is the civilian space agency of the U.S. government and was created in 1958 by an act of Congress. NASA's overall mission is to direct, plan, and conduct American civilian (including scientific) aeronautical and space activities for peaceful purposes. Available online. URL: http://www.nasa.gov/. Accessed on February 12, 2010.

**National Institute of Standards and Technology (NIST)** is an agency of the U.S. Department of Commerce that was founded in 1901 as the nation's first federal physical science research laboratory. The NIST Web site includes contemporary information about many areas of science and engineering, including analytical chemistry, atomic and molecular physics, biometrics, chemical and crystal structure, chemical kinetics, chemistry, construction, environmental data, fire, fluids, material properties, physics, and thermodynamics. Available online. URL: http://www.nist.gov/index.html. Accessed on February 12, 2010.

**National Oceanic and Atmospheric Administration (NOAA)** was established in 1970 as an agency within the U.S. Department of Commerce to ensure the safety of the general public from atmospheric phenomena and to provide the public with an understanding of Earth's environment and resources. Available online. URL: http://www.noaa.gov/. Accessed on February 12, 2010.

**NEWTON: Ask a Scientist** is an electronic community for science, math, and computer science educators and students sponsored by the Argonne National Laboratory (ANL) and the U.S. Department of Energy's Office of Science Education. This Web site provides access to a fascinating list of questions and answers involving the following disciplines/topics: astronomy, biology, botany, chemistry, computer science, Earth science, engineering, environmental science, general science, materials science, mathematics, molecular biology, physics, veterinary, weather, and zoology. Available online. URL: http://www.newton.dep.anl.gov/archive.htm. Accessed on February 12, 2010.

**Nobel Prizes in Chemistry and Physics.** This Web site contains an enormous amount of information about all the Nobel Prizes awarded in physics and chemistry, as well as complementary technical information. Available online. URL: http://nobelprize.org/. Accessed on February 12, 2010.

**Periodic Table of Elements.** An informative online periodic table of the elements maintained by the Chemistry Division of the Department of Energy's Los Alamos National Laboratory (LANL). Available online. URL: http://periodic.lanl.gov/. Accessed on February 12, 2010.

**PhET Interactive Simulations** is an ongoing effort by the University of Colorado at Boulder (under National Science Foundation sponsorship) to provide a comprehensive collection of simulations to enhance science learning. The major science categories include physics, chemistry, Earth science, and biology. Available online. URL: http://phet.colorado.edu/index.php. Accessed on February 12, 2010.

**ScienceNews** is the online version of the magazine of the Society for Science and the Public. Provides insights into the latest scientific achievements and discoveries. Especially useful are the categories Atom and Cosmos, Environment, Matter and Energy, Molecules, and Science and Society.

Available online. URL: http://www.sciencenews.org/. Accessed on February 12, 2010.

**The Society on Social Implications of Technology (SSIT)** of the Institute of Electrical and Electronics Engineers (IEEE) deals with such issues as the environmental, health, and safety implications of technology; engineering ethics; and the social issues related to telecommunications, information technology, and energy. Available online. URL: http://www.ieeessit.org/. Accessed on February 12, 2010.

**Spitzer Space Telescope (SST)** is a space-based NASA astronomical observatory that observes the universe in the infrared portion of the electromagnetic spectrum. This Web site contains contemporary information and educational materials about astronomy, astrophysics, and cosmology, including the infrared universe, star and planet formation, and infrared radiation. Available online. URL: http://www.spitzer.caltech.edu/. Accessed on February 12, 2010.

**Thomas Jefferson National Accelerator Facility (Jefferson Lab)** is a U.S. Department of Energy–sponsored laboratory that conducts basic research on the atomic nucleus at the quark level. The Web site includes basic information about the periodic table, particle physics, and quarks. Available online. URL: http://www.jlab.org/. Accessed on February 12, 2010.

**United States Geological Survey (USGS)** is the agency within the U.S. Department of the Interior that serves the nation by providing reliable scientific information needed to describe and understand Earth, minimize the loss of life and property from natural disasters, and manage water, biological, energy, and mineral resources. The USGS Web site is rich in science information, including the atmosphere and climate, Earth characteristics, ecology and environment, natural hazards, natural resources, oceans and coastlines, environmental issues, geologic processes, hydrologic processes, and water resources. Available online. URL: http://www.usgs.gov/. Accessed on February 12, 2010.

# Index

Note: *Italic* page numbers indicate illustrations; *m* indicates a map.

## D



## E

## X



## Y



## Z